Autorenkollektiv

Radicals in Biochemistry

Recent Developments with Copper Proteins

R. A. Lontie and D. R. Groeseneken

The Metabolism of Superoxide Anion and its Progeny in Blood Cells

R. E. Lynch

The Flavin Redox-System and its Biological Function

F. Müller

Semiquinone Formation in Flavo- and Metalloflavoproteins

D. E. Edmondson and G. Tollin

AKADEMIE-VERLAG · BERLIN

Autorenkollektiv

Radicals in Biochemistry

Autorenkollektiv

RADICALS IN BIOCHEMISTRY

With 13 Figures and 13 Tables

AKADEMIE-VERLAG · BERLIN
1983

Die Originalausgabe erscheint im Springer-Verlag Berlin · Heidelberg · New York

Vertrieb ausschließlich für die DDR und die sozialistischen Länder

Erschienen im Akademie-Verlag, DDR-1086 Berlin, Leipziger Straße 3—4
Lizenznummer: 202 · 100/529/83
Gesamtherstellung: VEB Druckerei „Thomas Müntzer", 5820 Bad Langensalza
Umschlaggestaltung: Karl Salzbrunn
Bestellnummer: 763 241 9 (6769) · 1275
Printed in GDR
DDR 68,— M

Table of Contents

Recent Developments with Copper Proteins

René A. Lontie, Dominique R. Groeseneken

Laboratorium voor Biochemie, Katholieke Universiteit te Leuven, B-3000 Louvain, Belgium

Table of Contents

Topics in Current Chemistry, Vol. 108
Managing Editor: Dr. F. L. Boschke

Abbreviations

BESOD: bovine erythrocyte (Cu,Zn)-superoxide dismutase
dopa: 3,4-dihydroxyphenylalanine
EC: Enzyme Commission
EPR: electron paramagnetic resonance
Hb: hemoglobin
M_r: relative molecular mass
NMR: nuclear magnetic resonance
PMN: polymorphonuclear neutrophils
SOD: superoxide dismutase, (Cu, Zn) when not stated otherwise

1 Introduction

Several copper containing oxidases and oxygenases avoid the liberation of free oxygen-radicals in the enzymatic reduction of dioxygen by their substrates, although a radical has been detected in some of them as an intermediate.

The copper proteins will be briefly reviewed. The intervention of radicals as substrate, possible intermediate, and reaction product of copper enzymes will be considered next. The pre-eminent role of the superoxide radical and of superoxide dismutase might make the whole review somewhat unbalanced. There have been in this field some hasty generalizations and there are several and sometimes lively controversies, like about the role of superoxide dismutases, the reaction mechanism and the half-site reactivity of the (Cu, Zn)-enzyme, its action on singlet oxygen, the non-photochemical production of singlet oxygen in biological systems, the mechanism of the Haber-Weiss reaction, the superoxide dismutase activity of ceruloplasmin, the state of copper in galactose oxidase.

Finally the action on copper proteins of mainly radiolytically produced radicals will be described. The role of the reactions of free radicals in biological systems in general is covered by comprehensive reviews [1,2].

2 Copper Proteins in General

The absorption of copper in the intestines and its storage in several organs in higher organisms have been linked with metallothioneins [3]. These proteins are characterized by M_r values of the order of 7000 and by a high cysteine content (nearly one third of the amino acids). In the liver e.g. they contain only a small amount of copper (about 1–3% of the metal content) besides small amounts of cadmium and very large amounts of zinc [4]. Other small proteins with an affinity for copper and with much lower cysteine contents have also been described [3].

The active sites of the main copper proteins can be classified according to the presence of mononuclear type-1 or (strongly) "blue" and type-2 or "non-blue" copper, or of binuclear type-3 copper [5].

The type-1 blue copper proteins act as electron carriers: azurin, plastocyanin, stellacyanin, umecyanin e.g. They are characterized by a rather strong LMCT (ligand to metal charge transfer) band near 600 nm and by small hyperfine coupling constants $A_{\parallel}$ in EPR. Copper is bound to two imidazole groups of histidine and to two sulfur atoms of methionine and/or cysteine in a slightly distorted tetrahedral environment, which seems inaccessible to water molecules [6-8].

The type-2 copper proteins, on the contrary, have a nearly square-planar copper environment, which is accessible to water molecules [9]. They show only weak absorption bands in the visible and normal $A_{\parallel}$ values in EPR. The main enzymatic activities are listed in Table 1.

The type-3 copper of hemocyanins [10] and tyrosinases [11] is usually EPR silent with the exception e.g. of *Helix pomatia* methemocyanin. Dopamine β-monooxygenase has been described to contain only type-2 copper according to EPR data [12], but there seems kinetic evidence for a transient binuclear site [13].

The "blue" oxidases contain these three types of copper together [6,14]. The stoichiometry is straightforward with laccase which contains one type-1 and one type-2 copper, and one type-3 dimeric copper site [6]. One would expect two laccase-like sites in ascorbate oxidase and in ceruloplasmin, but the presence of respectively 3 and 1 [15] and 1 and 3 [16] type-1 and type-2 copper atoms has been deduced. Ceruloplasmin shows oxidase activities towards different substrates, like Fe^{2+} (ferroxidase) and aromatic amines. It plays, moreover, an active role in the transport of copper [17]. With the proper precautions against the action of proteinases it can be isolated as a single polypeptide chain [18,19].

Cytochrome *c* oxidase contains one heme *a* and one heme a_3 besides two copper atoms: Cu_A EPR detectable when oxidized and Cu_B EPR undetectable [20-22]. With L-tryptophan 2,3-dioxygenase (EC 1.13.11.11), another heme protein, it has been

Table 1. Reduction of dioxygen by copper enzymes and their substrates (and cosubstrates)

Into	Enzyme	EC Number	Ref.
H_2O_2	Hexose oxidase	1.1.3.5	30)
	Galactose oxidase	1.1.3.9	31)
	Amine oxidase (copper containing)	1.4.3.6	32)
	Urate oxidase[a]	1.7.3.3	33)
	Lysyl oxidase[a]	1.13.12.aa	34)
H_2O	Cytochrome *c* oxidase	1.9.3.1	22)
	Catechol oxidase (tyrosinase)	1.10.3.1	11)
	[Monophenol monooxygenase (tyrosinase)	1.14.18.1]	
	Laccase	1.10.3.2	14)
	Ascorbate oxidase	1.10.3.3	35)
	Dopamine β-monooxygenase[b]	1.14.17.1	11, 12)
	Ceruloplasmin (ferroxidase)	1.16.3.1	17)

a Formation of H_2O_2 to be confirmed; b Cosubstrate: ascorbate shown that copper is not essential for its activity [23]. Similarly indoleamine 2,3-dioxygenase was devoid of copper [24], although the presence of two heme and two copper per M_r of 58,000 was reported simultaneously [25].

3 The Reduction of Dioxygen

The reactions of dioxygen have been amply documented [26–29]. The reduction can occur by a one-, two- or four-electron transfer reaction, the first of which is energetically unfavorable. The one-electron reduction of hydrogen peroxide yields the extremely reactive hydroxyl radical (Scheme 1).

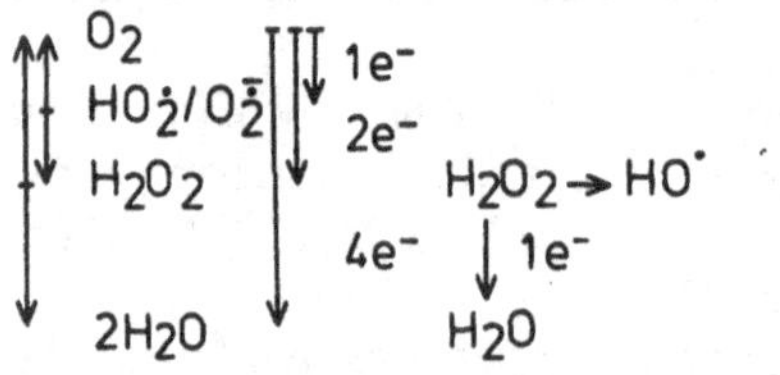

Scheme 1

3.1 The Reduction to Hydrogen Peroxide and to Water

A reduction to hydrogen peroxide is obtained with enzymes with one copper atom of type-2 (Table 1). A reduction to water requires enzymes with a dimeric site, with the three types of copper or with a four-metal center (2 heme, 2 Cu) (Table 1).

3.2 The Production of Superoxide Radicals in Biological Systems

An EPR signal, characteristic for the superoxide radical, was observed by the rapid-freezing technique in the oxidation at pH 10 of xanthine by dioxygen catalysed by xanthine oxidase (EC 1.2.3.2) [36]. The enzymatic reduction of dioxygen by aldehyde oxidase (EC 1.2.3.1) produces also the superoxide radical.

The plasma-membrane bound superoxide-generating flavoprotein (superoxide synthetase) of human polymorphonuclear neutrophils (PMN) has been solubilized by Triton X-100 [37] and deoxycholate [38] and purified. The enzyme catalyses the reaction:

$$2\,O_2 + NAD(P)H + H_2O \rightarrow NAD(P)^+ + 2\,O_2^{\cdot-} + H_3O^+ \tag{1}$$

The K_m value for NADPH is of the order of 35 μM against 900 μM for NADH. NAD(P)H seems to react on the cytosolic side of the membrane, while the superoxide anion is generated at the outer side [39].

Following a stimulation by the yeast cell wall extract zymosan, phagocytosing human leukocytes show an enhanced oxygen consumption, the cyanide-insensitive "respiratory burst", after a lag period of 30–40 s. The production of superoxide anions, measured by the reduction of Fe(III)-cytochrome *c*, followed the same course [40]. The formation of superoxide radicals was directly demonstrated by spin trapping on stimulated human neutrophils [41].

A similar enzymatic activity was found in stimulated pulmonar alveolar macrophages from humans [42], guinea pigs [43], and rats [44], and in retinal pigment epithelial cells from chick embryos [45].

A cyanide-sensitive NAD(P)H dependent superoxide generating system was described, however, for the nuclear membrane of hepatoma 22a ascites cells grown in mice [46].

The formation of superoxide has, moreover, been reported for cytochrome *P*-450, diamine oxidase, flavoproteins, and peroxidases [47].

Superoxide radicals can also be produced in the reduction of dioxygen by strong reducing agents, like iron-sulfur proteins, flavins, phenylhydrazine, semiquinones, and thiols [47-49]. Some of the reactions of dioxygen, like with thiols, appear to be metal catalysed *in vitro* [50]. Several substances, when reduced intracellularly, are able to generate superoxide anions with dioxygen, like the herbicide paraquat (1,1′-dimethyl-4,4′-bipyridinium dichloride) in *Escherichia coli* [51] and with mouse lung microsomes [52].

The production of superoxide radicals was also observed in the "autoxidation" of oxyhemoglobin to methemoglobin: with shark hemoglobin by the induction of the oxidation of adrenaline to adrenochrome [53] and with bovine hemoglobin, in 1 M phosphate buffer pH 6.0, by the reduction of Fe(III)-cytochrome *c* [54]. The "autoxidation" of bovine oxymyoglobin to metmyoglobin also induced the oxidation of adrenaline to adrenochrome [55]. A displacement of a superoxide anion from the active site by a water molecule or hydroxyl anion in a proton-catalysed process follows from a kinetic investigation of the influence of pH and temperature on the "autoxidation" of bovine myoglobin [56]. Similarly with human hemoglobin a proton-assisted displacement of superoxide by nucleophilic anions has been described [57], although the authors have considerably modified their first interpretation [58] (see Sect. 4.1.7).

Finally superoxide radicals can also be generated photochemically in chloroplasts in the presence of ascorbate or of paraquat. The formation was demonstrated by spin trapping on illumination of spinach chloroplasts in the presence of oxygen and paraquat [59]. Superoxide radicals are formed, moreover, in the near-ultraviolet photooxidation of tryptophan, as indicated by the increase of the H_2O_2 production in the presence of SOD [60], and on irradiation in aerated solutions of protoporphyrin at 400 nm [61] and of melanin with light of 320–600 nm [62], as shown by spin-trapping.

In aprotic solvents the superoxide anion is a very weak oxidant: the uptake of a proton from the substrate is followed by the dismutation of $HO_2^{\cdot}$ [63]. Thus stable solutions of $O_2^{\cdot}$ can be prepared by controlled-potential coulometry [63,64].

3.3 The Production of Hydroxyl Radicals in Biological Systems

Although aqueous superoxide is a potential fairly strong oxidant $E^{0\prime}(O_2^{\cdot-}/H_2O_2) = 0.94$ V at pH 7.0 [65], it is not considered very reactive, except e.g. with ascorbate and with NADH bound to lactate dehydrogenase [66], and with the thiol group of cysteine, as shown by the inactivation of papain [67].

The production of ethylene from methional (3-thiomethylpropanal) was induced by the oxidation of xanthine by dioxygen catalysed by xanthine oxidase [68]. The second-order rate constant for the reaction of hydroxyl radicals with methional was estimated by pulse radiolysis [69] to amount to $8.2 \times 10^9\ M^{-1}\ s^{-1}$, while the superoxide anion reacted more slowly [69,70]. The short lag period of the ethylene production induced by the oxidation of xanthine could be overcome by the addition of small amounts of hydrogen peroxide. The reaction was inhibited by SOD or by catalase, and by scavengers of hydroxyl radicals, so that the Haber-Weiss reaction was implicated [68]:

$$H_2O_2 + O_2^{\cdot-} \rightarrow O_2 + {}^{\cdot}OH + OH^- \tag{2}$$

A direct Haber-Weiss reaction is generally considered to be far too slow [71,72]; a second-order rate constant of $2.3\ M^{-1}\ s^{-1}$ has been estimated from the decomposition of aqueous H_2O_2 by γ rays [73]. The reaction could be metal catalysed by a Fenton-type mechanism [74,75]:

$$Fe^{3+}\text{-chelate} + O_2^{\cdot-} \rightarrow O_2 + Fe^{2+}\text{-chelate} \tag{3}$$

$$H_2O_2 + Fe^{2+}\text{-chelate} \rightarrow Fe^{3+}\text{-chelate} + {}^{\cdot}OH + OH^- \tag{4}$$

The Fe-EDTA complex catalysed the oxidation of tryptophan [76] and the hydroxylation of salicylate [77] in the presence of a superoxide-generating system. A partial inhibition by diethylenetriaminepenta-acetate was observed for the latter reaction [77] and in the ethylene production from 2-keto-4-thiomethylbutyrate [75].

The production of hydroxyl radicals was demonstrated by spin-trapping in the autoxidation at pH 11.5 of 6-hydroxydopamine. The reaction was partially inhibited by SOD, by catalase, and by diethylenetriaminepenta-acetate and deferoxamine [78].

A reduction and activation of H_2O_2 by other one-electron donors, like semiquinones, has also to be considered. This follows from a study of the ethylene production from methionine in the presence of pyridoxal phosphate, a reaction characteristic for ${}^{\cdot}OH$ radicals or for Fenton-type oxidants. The ethylene production in the presence of dioxygen, anthraquinone-2-sulfonate, and an NADPH-generating system in phosphate buffer pH 7.6 was inhibited by SOD and by catalase, but stimulated by scavengers of ${}^{\cdot}OH$ radicals, like 0.1 mM mannitol, α-tocopherol, and formiate [79].

The production of hydroxyl radicals was also shown by the oxidation of benzoic acid, a specific scavenger, with stimulated human granulocytes [80] and by the formation

Table 2. N- & C-Terminal amino-acids and conserved amino-acids at the active site of SOD

Species	N-term.	His	His	His	His	His	Asp	His	Arg	C-term.	Ref.
Yeast[a]	Val	46	48	63	71	80	83	120	143	Asn-153	99, 100)
Bovine[b]	Ac-Ala	44	46	61	69	78	81	118	141	Lys-151	101)
Equine[c]	Ac-Ala	46	48	63	71	80	83	120	143	Phe-153	102)
Human[b]	Ac-Ala	46	48	63	71	80	83	120	143	Gln-153	103, 104)

a *Saccharomyces cerevisiae*; b Erythrocytes; c Liver

of ethylene from methional with stimulated human alveolar macrophages [81]. For the detection by spin-trapping of the hydroxyl-radical production by stimulated human neutrophils [41] an alternate explanation was suggested: the decomposition of the superoxide adduct of the spin-trap [82].

4 Superoxide Radicals as Substrate

4.1 (Cu,Zn)-Superoxide Dismutase

4.1.1 Discovery

Hemocuprein has already been crystallized in 1938 from bovine erythrocytes [83]. It contained 0.34% Cu and was also present in small amounts in blood serum. Similar proteins have been isolated from the erythrocytes of other mammals and from tissues like liver and brain and named accordingly: erythrocuprein (human erythrocytes) [84], hepatocuprein (bovine and equine liver), cerebrocuprein (human brain). All these proteins were called cytocupreins, as they showed the same Cu content and M_r value, a weak absorption in the visible, a lack of enzymatic activity, and for a given species an immunological identity. These cytocupreins were found, moreover, to contain 2 mol of Zn besides 2 mol of Cu per M_r of 33,600 [85].

The inhibition by erythrocuprein or hemocuprein of the reduction of Fe(III)-cytochrome *c* by a reaction product of xanthine oxidase — the superoxide radical — indicated the superoxide dismutase activity (EC 1.15.1.1) of the cytocupreins [86].

Two symposia have entirely been devoted to superoxide dismutases [87,88] like several reviews [47,49,89–95]. Much information can also be found in three books [96–98].

4.1.2 Sequence and Conformation

The amino-acid sequence has been determined for the SOD of four species (Table 2). There are minor differences for the yeast enzyme: Asn-55 and Asn-92 [99] against Asp-55 and Asp-92 [100], and for the human enzyme concerning the amidation of six residues and: Ser-17 and Val-98 [103] against Ile-17 and Ser-98 [104]. Characteristic is the very low amount or the absence of methionine and tryptophan. There is no methionine in human SOD, the yeast SOD contains only Met-84, the bovine SOD Met-115, and the equine SOD Met-99 and Met-117. Tryptophan is absent with the exception of Trp-32 in human SOD.

A sequence homology of 54% was observed between the yeast and bovine SOD [99], of 80% between equine and bovine SOD [102], and of 82% between the bovine and human SOD [103].

The extra cysteine residue (Cys-111) of human SOD [103,104] is responsible for the formation of a mixed disulfide [105] or of R—S—$(S)_n$—R′ ($n \geq 0$) [106] as an artefact of the preparation, yielding the absorption band at 325 nm [107], which is not formed in the SOD of other species investigated so far.

The conformation of BESOD has been determined by X-ray diffraction to a resolution of 0.3 nm [108] and of 0.2 nm, as quoted in [109]. Almost 50% of the amino-acid residues form a β barrel made up of 8 antiparallel strands. There is a striking similarity with the folding pattern of the domains in immunoglobulins. Cys-55 and Cys-144 form a disulfide bridge. The ligands of Cu and Zn have been identified, the metal atoms are at a distance of 0.6 nm with the bridging imidazolate anion of His-61 (Scheme 2). Zn(II) is in a nearly tetrahedral and Cu(II) in a slightly distorted square-planar environment. It was shown that BESOD contains two paired subdomains [110].

```
        His-118              His-69
His-46   (Cu)   His-61   (Zn)   His-78
        His-44               Asp-81
```

Scheme 2

The ligands of Cu and Zn are conserved in the sequences (Table 2). High resolution ^{1}H NMR showed that the active sites of yeast and bovine SOD have the same geometry [111]. A great structural homology was similarly established between the active sites of human and bovine SOD [112] and of BESOD and the two SOD isoenzymes of wheat germ [113]. The ligand histidine resonances were identified from the exchange of the histidine C-2 protons, followed by ^{1}H NMR spectroscopy at 270 MHz. The method was applied to yeast and bovine SOD by comparing the reduced, the Cu-free, and the metal-free proteins [114].

4.1.3 Quaternary Structure and Half-Site Reactivity

SOD occurs as a homodimer in solution: M_r about 33,000 [84, 107], twice the value from the sequence. The two active sites are identical as shown by X-ray diffraction [108].

The rapid-freezing EPR data obtained in pulse radiolysis experiments had been interpreted by half-site reactivity [115]. In the meantime preparations have, however, been obtained with a definitely higher specific activity (second-order rate constant 3.3×10^9 against 2.4×10^9 M^{-1} s^{-1} for the preparations used in the pulse radiolysis experiments) [116], so that the half-site reactivity has been seriously questioned [117] by the presence of inactive enzyme in the earlier experiments [115]. Large differences in specific activity (up to a factor of two) and in the ability to protect mouse bone marrow stem cells against X-irradiation have been reported for five SOD preparations [118].

New arguments in favor of an interaction between the Cu sites in the dimer have nevertheless been advanced. The binding of Cu^{2+} to a first site of BESOD seems to facilitate the binding to the second site [119]. On reconstitution of SOD by the addition of Cu^{2+} to a Cu-free preparation, while the recovered spectral properties were a linear function of the Cu content, the recovery of the activity was interpreted by a specific activity of the dimer with one Cu which was twice that of the dimer with two Cu [120].

The SOD of the swordfish *Xiphias gladius* remained fully active in 8 M urea. It was concluded from gel chromatography that active monomers were present [121]. But the M_r value in 8 M urea of the SOD, presumably very stable in this medium, was obtained with standard proteins, considered unstable under these conditions. Bovine and yeast SOD showed also an apparent M_r value of 16,000 in 8 M urea by gel chromatography after an abrupt decrease between 6 and 7 M urea. BESOD retained its full enzymatic activity, while yeast SOD was reversibly inactivated [122].

The extent of the dissociation in 8 M urea can, moreover, not be estimated from the electrophoretically observed hybrid formation [122, 123] between wheat germ (isoenzyme II) SOD and respectively yeast and bovine SOD. BESOD did namely not dissociate in 8 M urea at 25 °C for at least 72 h, as indicated by sedimentation equilibrium analysis (M_r 32,000). The reactivity of Cys-6 and of the histidine residues was not increased either in that medium [124].

4.1.4 Reconstitution, Preparation and Investigation of Metal Derivatives

In stating the metal content of the enzyme and of its metal derivatives only one subunit will be considered. The Cu site is presented to the left and the Zn site to the right: (Cu,Zn) e.g. for the native enzyme.

Cu^{2+} and Zn^{2+} can entirely be removed from BESOD by chromatography on a Sephadex G-25 column by elution with 10 mM EDTA at pH 3.8 [125]. At this pH the apoprotein is stable and a total reconstitution was achieved by the stoichiometric addition of Cu^{2+} and subsequently of Zn^{2+}, followed by a dialysis against phosphate buffer pH 7.8. Zn^{2+} could also be replaced by Co^{2+}, Cd^{2+}, and Hg^{2+}, yielding active products. The (Cu,—)-derivative with an empty Zn site was only partially active [126]. It is important in such experiments to remove EDTA entirely from the apoprotein, like by a dialysis against 0.1 M $NaClO_4$ [127].

The binding of one Zn^{2+}, presumably at the Zn site, organizes the active site, as shown with BESOD by ^{1}H NMR spectroscopy [128, 129] and by chemical modification studies with diethylpyrocarbonate [128]. The ligands of the Zn site are close together in the sequence (Scheme 2).

A very interesting (Cu^{II},Cu^{II})-derivative of BESOD was prepared with 2 Cu^{2+} per subunit. The EPR spectrum between 5 and 10 K indicated an antiferromagnetic interaction of the two Cu^{2+} with a coupling constant of 52 cm^{-1}, which seems very good evidence for an imidazolate-anion bridge [130], as borne out with model complexes [94]. Similarly the (Cu^{II},Co^{II})-derivative of BESOD showed only a weak EPR spectrum [126, 127, 131]. On reduction of Cu(II) to Cu(I) a normal EPR spectrum of Co(II) was observed at liquid-helium temperatures like with the (—,Co^{II})-derivative [127]. Magnetic susceptibility measurements over a wide range of temperatures yielded an antiferromagnetic coupling constant $2J \gtrsim 600$ cm^{-1} for the (Cu^{II},Co^{II})-BESOD [132].

In phosphate buffer pH 7.4 a (Co^{II},Co^{II})-derivative of BESOD was obtained, while the addition of Co^{2+} in acetate buffer pH 5.4 yielded the (—,Co^{II})-derivative [133].

The inactive (Ag^{I},Cu^{II})- and (Ag^{I},Co^{II})-derivatives of BESOD have also been prepared. The EPR spectra of (Ag^{I},Cu^{II})- and (Cu^{I},Cu^{II})-BESOD were very similar like the visible absorption spectra of (Ag^{I},Co^{II})- and (Cu^{I},Co^{II})-BESOD. From model fitting to the electron-density map at 0.3-nm resolution by an interactive computer graphics system is was concluded that the bridging His-61 is slightly out of the plane of Cu and its three other histidine ligands [134].

The metal ion is lost reversibly from the Zn-binding site of BESOD below pH 4. With the (Cu^{II},Cu^{II})-derivative an EPR spectrum at $g = 2$ appeared reversibly, suggesting the loss of Cu^{2+} from the Zn-binding site [135]. With the Zn-free (Cu^{II},—)-BESOD a reversible migration of Cu^{2+} from one subunit to the Zn site of another was observed to occur with a pK_a of 8.2 by the decrease of the EPR signal at 30 °C [136].

The pronounced increase above pH 9.3 of the proton nuclear-magnetic relaxation rates in aqueous solutions of (Cu^{II},Zn^{II})-BESOD at 295 K and 30 MHz was interpreted by the cleavage of His-61 from Cu^{2+} and its replacement by an equatorial OH^-. The axial H_2O on Cu^{2+} remained unaffected [137].

The metal-substituted derivatives of BESOD and the influence of pH on the reconstitution and on some of the derivatives have been thoroughly discussed [138].

With the yeast enzyme it was concluded that His-63 is not a bridging ligand from the perturbed angular correlation of γ rays, which allows to detect the nuclear quadrupole interaction between a ^{111}Cd nucleus and the environment. The (Cu^{II},Cd^{II})- and $(—,Cd^{II})$-derivatives showed namely the same spectrum, due to the presence of two species. With the (Cu^{I},Cd^{II})-derivative, which showed a different spectrum due to only one species, it was concluded, moreover, from the similarity of the nuclear quadrupole interaction parameters with those of ^{111}Cd-carboxypeptidase A at low pH that His-63 is apparently not a ligand of Cd(II) [139]. The interpretation of the ^{3}H exchange rate of the histidine C-2 protons in yeast (Cu^{II},Zn^{II})-SOD was straightforward for the ligands of Cu (His-46, His-48, His-120) and for two ligands of Zn (His-71, His-80). From its higher incorporation rate it was inferred that His-63 is not coordinated to Cu, but only to Zn [140]. Small differences between the active sites of BESOD and of the yeast enzyme are plausible, as indicated already by the reversible inactivation of the latter in 8 M urea [122]. But one would like to have the same experimental data with $(Cu,^{111}Cd)$-BESOD like EPR or magnetic susceptibility data for the (Cu^{II},Cu^{II})-derivative e.g. of yeast SOD.

4.1.5 Reaction Mechanism

An alternate reduction and oxidation of Cu by superoxide anions (Scheme 3) followed from pulse radiolysis experiments [115, 141]. The $E^{0'}$ value (pH 7.0) of Cu in BESOD was estimated to be 0.42 V [142]. In the presence of tetranitromethane, an effective scavenger of superoxide radicals, the dismutation reaction was reversed, as BESOD catalysed the reduction of dioxygen by hydrogen peroxide [143].

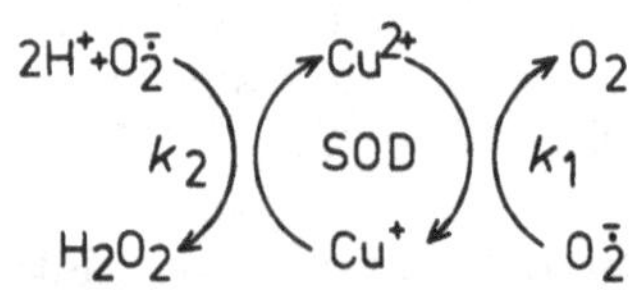

Scheme 3

Cu^{2+} in BESOD is slowly and partially reduced by H_2O_2 [141, 144–147] with a second-order rate constant of 6.7 M^{-1} s^{-1} at pH 10.0 and 25 °C [147]. The formation of the superoxide anion in this reaction seems thermodynamically unfavorable, but the rapid dismutation of 2 $O_2^{\cdot-}$ to O_2 and H_2O_2 will make the overall reaction possible [142]. A secondary reaction with the destruction of one histidine per subunit leads to the inactivation of SOD. It was explained by a Fenton-type mechanism between H_2O_2 and Cu(I) in the active site [146, 147] and is accompanied by chemiluminescence [148]. This secondary reaction must obviously be avoided in kinetic experiments.

The X-ray absorption edge brought direct evidence for the presence of Cu(I) in BESOD reduced by dithionite, while the absorption edge of Zn^{2+} was unaffected, as expected. The action of H_2O_2 was incomplete and shifted also the absorption edge to lower energies. The Cu(I) environment in SOD reduced with dithionite was very similar to that in [Cu(I)(imidazole)$_4$], while there was a difference between Cu^{2+} in BESOD and in [Cu(II)(imidazole)$_4$]. Upon reduction the environment of Cu from square planar became more tetrahedral [149].

The reduced (Cu^{I},Zn^{II})-BESOD is only very slowly reoxidized by dioxygen [141, 142, 145, 146].

The dismutation of $O_2^{\overline{\cdot}}$ catalysed by BESOD occurs with a second-order rate constant under turnover conditions of up to 3.7×10^9 M^{-1} s^{-1}, near the limit for a diffusion-controlled reaction, in 0.5 mM sodium pyrophosphate, 100 μM EDTA, 0.1 M ethanol at pH 9.0, where the nonenzymic decay of superoxide is slow [116]. The approach of the steady state was followed in pulse-radiolysis experiments by measuring transmittance changes at 650 nm with native and with (partially) reduced BESOD [115]. The rate constants for the reduction of (Cu^{II},Zn^{II})-SOD and for the reoxidation of (Cu^{I},Zn^{II})-SOD corresponded with the turnover rate.

Somewhat lower values of 1.0×10^8 and 1.2×10^8 M^{-1} s^{-1} at pH 7.0 and 8.5 respectively were deduced from a kinetic study of the inhibition by green pea SOD of the reduction of Fe(III)-cytochrome *c* with xanthine oxidase as a superoxide generating system [150].

The rate constant ($k_1 = k_2$ in Scheme 3) is independent of pH between pH 5.3 to 9.5 [151]. In order to provide a proton to O_2^{2-} formed in the reduction of $O_2^{\overline{\cdot}}$ it was suggested that the imidazolate anion of His-61 in BESOD is detached from Cu upon reduction to Cu(I) and protonated [147]. From the variation with pH of the midpoint potential of Cu in BESOD it had been shown that the reduction of Cu^{2+} is accompanied between pH 5 and 8.7 by the uptake of a proton [142]. An open coordination on one of the metal atoms was deduced for BESOD reduced with dithionite from the contribution to the relaxation of the nucleus of $^{35}Cl^-$ anions [152]. It is presumably located on Cu(I) as the (—,Co^{II})- and (Cu^{I},Co^{II})-derivatives of BESOD showed the same absorption spectrum [153, 154] and as the X-ray absorption edge of Zn^{2+} was not modified upon reduction of Cu^{2+} with dithionite [149]. From the similarity of the NMR spectra of the (—,$^{113}Cd^{II}$)- and (Cu^{I},$^{113}Cd^{II}$)-derivatives of BESOD it was concluded that His-61 remains attached to Cd upon reduction of Cu^{2+} and that the imidazolate anion is protonated on the Cu side [155]. Rather different results have been presented for $^{113}Cd^{2+}$-substituted BESOD [156], which were discussed in [155].

Is such an open coordination on Cu(I) in BESOD, presumably occupied by a water molecule [152], entirely compatible with the mentioned similarity of the X-ray absorption edge of reduced (Cu^{I},Zn^{II})-BESOD and Cu^{I}(imidazole)$_4$ [149]? A distinction between O of water and N of imidazole might not have been possible.

The spectral changes observed with (Cu^{II},Co^{II})- and (Cu^{I},Co^{II})-BESOD respectively by reduction and oxidation with $O_2^{\overline{\cdot}}$ under turnover conditions were interpreted by a protonation and deprotonation of the bridging imidazolate anion. They were, moreover, shown to occur at the rate of the reduction and oxidation of the Cu center [154].

An essential arginine is also conserved in the sequences (Table 2). By blocking arginine-141 of BESOD or of the homologous residue of SOD of several other species with butanedione in borate buffer the activity was lost for about 90% [157]. SOD from baker's yeast was inactivated for more than 98% on blocking arginine-143 [158]. With (Cu^{I},Zn^{II})- and (Cu^{II},Zn^{II})-SOD from yeast ^3H was incorporated into a peptide which contained arginine-143 [140]. The binding of the substrate at the Cu center and the transfer of a proton to the generated peroxide dianion have been suggested as a double role for this essential arginine [157].

An attractive mechanism takes moreover into account the hydration of $O_2^{\overline{\cdot}}$ and the structure of water in the active site [159]. The reduction of Cu^{2+} in SOD by $O_2^{\overline{\cdot}}$ is supposed to occur through the axial H_2O, which was shown to exchange only very slowly [137]. Scheme 4 incorporates this mechanism [159] together with the protonation-deprotonation of the bridging imidazolate [147, 155].

A mechanism without reduction of Cu^{2+}, similar to the reaction between $HO_2^{\cdot}$ and $O_2^{\overline{\cdot}}$, was presented for the catalysis of the dismutation of superoxide by the Cu^{2+} cation [160]:

$$Cu^{2+} + O_2^{\overline{\cdot}} \rightarrow CuO_2^{+} \tag{5}$$

$$CuO_2^{+} + O_2^{\overline{\cdot}} + 2\,H_2O \rightarrow Cu^{2+} + O_2 + H_2O_2 + 2\,OH^- \,. \tag{6}$$

Scheme 4

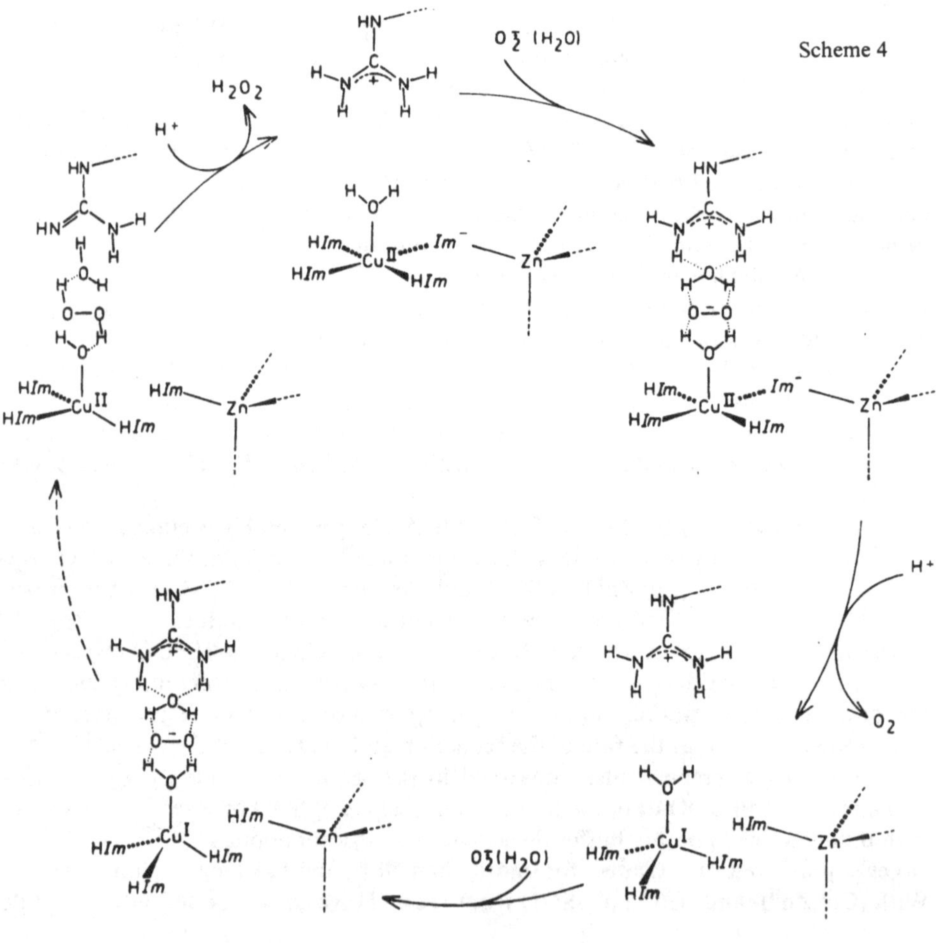

The formation of the analogous intermediate MnO_2^+ was demonstrated in the reaction of the Mn^{2+} cation with the superoxide radical [161].

Aqueous solutions of SOD from human placenta were treated with γ rays at 77 K and annealed at 190 K. An EPR signal observed at 77 K was interpreted as due to a paramagnetic intermediate: a "loose complex" Cu^{2+} ... $HO_2^{\cdot}$ at the active site [162]. The binding of $HO_2^{\cdot}$ ($pK_a = 4.9$) instead of $O_2^{\cdot-}$ seems rather improbable in phosphate buffer pH 7.8. The activity of BESOD is, moreover, independent of pH above pH 5.3 [151]. The paramagnetic intermediate could correspond to a superoxide radical bound to the axial H_2O of Cu^{2+} (Scheme 4).

The decay near 200 K of the EPR signal of $O_2^{\cdot-}$ in the "loose complex" of BESOD was not accompanied by a reduction of Cu^{2+}, which occurred only upon annealing at about 270 K. Similar observations were made with BESOD reduced with H_2O_2. This was considered good evidence for the second mechanism (5) (6) with a constant oxidation state of Cu, Cu^{2+} or Cu^+, and a cooperation of the two active centers in the SOD dimer [163].

Such a cooperation seems very unlikely, the active centers being at a distance of 3.4 nm [108]. With BESOD a heterodimer was prepared with a native subunit and a diazo-coupled H_2O_2-inactivated subunit, whereby the activity was not altered. The heterodimer showed approximately half the specific activity of the native enzyme [164], which excludes moreover the half-site reactivity [115, 117] (See Sect. 4.1.3). These results were confirmed with the (Cu,Zn) (—,Zn)- and (Cu,Zn) (Co,Zn)-dimers, which showed almost the same specific activity, expressed per Cu, as the native enzyme [165].

From the liver of rats treated with lethal doses of X rays an inactive, presumably reduced, enzyme was isolated, which was only very slowly reactivated in air [166]. This derivative must obviously differ from the assumed (Cu^I,Zn^{II})-intermediate in the catalytic process (Scheme 4).

4.1.6 Distribution

The (Cu,Zn)-SOD is found in the cytosol of eukaryotic cells and in the intermembrane space of mitochondria, as observed with chicken liver [167], and also in chloroplasts, as shown with spinach [168] and mustard [169] leaves. A (Cu,Zn)-SOD has also been isolated from a prokaryote *Photobacterium leiognathi* [170]. Could this symbiotic bacterium have borrowed the gene(s) from its host, the Pony fish, although two different subunits (M_r 15,000 and 17,000) were reported for the bacterial enzyme? A (Cu,Zn)-SOD was furthermore detected in *Paracoccus denitrificans* [171].

Fe- and Mn-containing superoxide dismutases have, moreover, been isolated. They show striking sequence homologies. (Fe)-SOD has mainly been found in prokaryotes. (Mn)-SOD occurs in prokaryotes, M_r $(20{,}000)_2$, in the matrix of mitochondria, although encoded by the nucleus, and also in the cytosol of e.g. liver cells (human, chicken in contrast with rat) [172, 173].

Superoxide dismutases are found in all aerobic and in aerotolerant organisms, which consume oxygen [174], with the exception of *Mycoplasma pneumoniae* and other species of *Mycoplasma* [175].

The three types of SOD showed a similar activity near pH 7.0 with a second-order rate constant of about $2 \times 10^9\ M^{-1}\ s^{-1}$, as determined by the reduction of Fe(III)-cytochrome *c* and of tetranitromethane. The activity of the (Fe)- and (Mn)-SOD decreased with increasing pH in contrast with that of (Cu,Zn)-SOD [176].

The addition of 1 mM cyanide, which inhibits the (Cu,Zn)-SOD, allows to distinguish it from (Mn)-SOD in samples of eukaryotic origin [177,178]. With azide a 50% inhibition was observed at pH 7.8 for (Cu,Zn)-, (Mn)-, and (Fe)-SOD at a concentration of about 32, 20, and 4 mM respectively [179]. Pamoate, however, was only an apparent inhibitor of the three types of SOD, as it lowered the sensitivity of several assays [180].

In rat liver 16% of the dismutase activity was found in the mitochondria: 9% in the intermembrane space and 7% in the matrix, due respectively to (Cu,Zn)- and (Mn)-SOD. No activity was located in other subcellular structures [181].

A radioimmunoassay has been developed for the (Cu,Zn)-SOD [182] and a ^{19}F nuclear magnetic relaxation method for the (Cu,Zn)- and (Mn)-SOD [183].

The gene for human (Cu,Zn)-SOD is located on chromosome 21 (two alleles have actually been described) [184]. In trisomy 21 (Down syndrome) an increase of about 50% in SOD activity was observed in erythrocytes and in blood platelets [184,185].

As demonstrated with ^{67}Cu in rats, (Cu,Zn)-SOD receives its copper from ceruloplasmin after 2–3 days [186]. The (Cu,Zn)-SOD activity in erythrocytes is reduced in Cu deficiency, as shown with several species [187]. With rats e.g. during Cu depletion, plasma Cu and ceruloplasmin were decreased by 78 and 75% respectively against 72% and only 56% for the blood cell Cu and (Cu,Zn)-SOD respectively [188]. In three patients with Wilson's disease the SOD level of erythrocytes was normal, although the disease is characterized by an accumulation of Cu in the liver e.g. and usually by low concentrations and sometimes the absence of ceruloplasmin [189].

The (Cu,Zn)-SOD levels in erythrocytes of normal human subjects are fairly constant: 461 $\pm$ 46 (standard deviation) µg/g hemoglobin. Results are also reported for several pathological conditions [185]. Small differences were e.g. observed in erythrocytes with on the average a 19% decrease in patients with Duchenne muscular dystrophy [190] and a 20% increase in black alcoholics in contrast with white alcoholics [191]. Sickled erythrocytes showed significantly higher levels of (Cu,Zn)-SOD: 170% on the average of those of normal erythrocytes [192]. Tumor cells as a rule seem to have lower or zero levels of (Mn)-SOD [193].

The liver (Cu,Zn)-SOD of old rats (27 months) showed a 60% reduction of specific activity, the presence of antigenically cross-reacting material, and a decreased thermal stability in comparison with the enzyme of young rats (6 months) [194]. Similar observations were made with ageing human fibroblasts in culture, whereby the absence of a cytosolic stabilizing factor was suspected in old cells [195].

4.1.7 Role and Applications

The biological role and the clinical applications of superoxide dismutases are covered by the second volume of the Symposium in Malta [196].

Superoxide dismutases together with catalase (EC 1.11.1.6) and glutathione peroxidase (EC 1.11.1.9) are important enzymes for the protection against "oxygen toxicity" [91,197–199]. Protection is also provided by substances like α-tocopherol (vitamin E) [200].

A synergism between SOD and catalase points to the $^{\cdot}OH$ radical (or a $^{\cdot}OH$ radical like intermediate) as the toxic species. The damage to resealed erythrocyte ghosts by radiolytically produced $^{\cdot}OH$ radicals was enhanced by H_2O_2 by a factor of almost two. This could be due to the formation of a second ·OH radical by the reaction of H_2O_2 with a radical formed from the substrate in the first step [201].

The role of superoxide dismutase has been challenged [47,202]. One argument was their presence in anaerobic bacteria, although often in low amounts [203]. Several anaerobes do also contain catalase. A defense against an accidental exposure to dioxygen has been suggested as a possible explanation for the presence of both enzymes in aerotolerant anaerobes [203–205].

A further role has been attributed to (Cu,Zn)-SOD: the quenching of singlet oxygen. This raises, moreover, the question of the unlikely production of singlet oxygen in biological systems by non-photochemical means.

It has been suggested that the spontaneous dismutation of $O_2^{\overline{\cdot}}$ produces singlet oxygen in contrast with the SOD-catalysed dismutation [206]:

$$HO_2^{\cdot} + O_2^{\overline{\cdot}} + H_3O^+ \rightarrow H_2O_2 + H_2O + {}^1O_2 \,. \tag{7}$$

A quenching by SOD of singlet to triplet oxygen has been furthermore advocated [207–209].

No singlet oxygen was, however, found in the reaction of $O_2^{\overline{\cdot}}$ with one-electron acceptors [210] and no [211,212] or no more than 0.2% [213] in the spontaneous dismutation of $O_2^{\overline{\cdot}}$. No quenching by SOD of singlet oxygen was observed either [214–216]. The weak luminescence in the aerobic xanthine oxidase reaction was inhibited by SOD, by catalase, and by scavengers of ·OH radicals. Attributed to singlet oxygen, it seems due to carbonate radicals, formed by the action of ·OH radicals on carbonate ions [217]. Similarly there is no good evidence for the production of singlet oxygen by the myeloperoxidase of neutrophils. HOCl, produced from Cl^- anions and H_2O_2, reacts with the singlet-oxygen traps [218–220]. The chemiluminescence of phagocytosing neutrophils, also attributed to singlet oxygen, seems to be mainly due to the oxidation of ingested particles [221].

Intravenously administered (Cu,Zn)-BESOD, when coupled to high molecular mass derivatives in order to keep it in the circulation, showed a marked anti-inflammatory effect [222]. The anti-inflammatory protein "orgotein", isolated from a bovine liver extract in 1965 and which contained both Cu and Zn, was found to be identical with (Cu,Zn)-SOD [223–224].

Two elements in inflammatory processes are the attraction of leukocytes and the vasodilatation caused by prostaglandin E_1. Superoxide radicals generate in plasma a factor chemotactic for neutrophils [225]. The prostaglandin phase of carrageenan induced foot edema in rats was suppressed by frequent intravenous injections of BESOD. Heat-denatured BESOD, catalase, and ·OH-radical scavengers were inactive [226]. The granuloma formation in rats, also induced by carrageenan-soaked sponges and measured after 7 days, was not inhibited by a local injection of BESOD alone at the moment of sponge implantation, but by a combination of BESOD and catalase, which was possibly due to a protection of the former enzyme by the latter [227].

Hyaluronate of bovine synovial fluid was extensively degraded within 30 min in the presence of the xanthine oxidase system, as indicated by the decrease of the relative viscosity. This degradation of hyaluronate was inhibited by (Cu,Zn)-SOD and by catalase [228]. The formation of ·OH radicals was attributed to an iron-catalysed

Haber-Weiss reaction [229]. Hyaluronate of the bovine vitreous body was similarly protected by (Cu,Zn)-SOD, by catalase, and by peroxidase. The degradation of hyaluronate in the presence of ascorbate or of Fe^{2+} cations was not inhibited by (Cu,Zn)-SOD [230]. The formation of $^{\cdot}OH$ radicals from H_2O_2 and ascorbate, in the presence of traces of Fe-EDTA, was largely independent of superoxide radicals [231].

Rats, exposed for 7 days to 85% O_2, survived for more than 4 days in 100% O_2, when control rats died within 3 days. They showed a 50% increase of SOD in their lungs. With guinea-pigs, hamsters, and mice there was no increase of pulmonary SOD and no tolerance was developed [232]. This increase in pulmonary SOD at 80% O_2 or more was only observed with 10-day old and not with 25-day old rats. Catalase and glutathione peroxidase were increased with both groups [233]. There are, however, modifications at the cellular level besides the increase of the enzyme activities.

Rabbit peritoneal macrophages ($p_{O_2} \simeq 15$ mmHg) showed only half the SOD activity, expressed per mg protein, of the alveolar macrophages ($p_{O_2} \simeq 150$ mmHg). Lung and brain SOD activities of hypoxic mice were also significantly lower than in normoxic controls [234].

A correlation was also presented between life-span and (Cu,Zn)- plus (Mn)-SOD activities of two strains of *Drosophila melanogaster*. The short-living vestigial strain had an activity of 72% on the average of the wild strain [235]. With 2 rodent and 12 primate species there was a good correlation between the ratio of SOD activity to specific metabolic rate and the maximum life-span potential for the brain, heart, and liver [236].

The "autoxidation" of oxyhemoglobin, like that of oxymyoglobin, produces $O_2^{\overline{\cdot}}$ (See Sect. 3.2):

$$HbFe^{3+}—O_2^{\overline{\cdot}} \rightarrow HbFe^{3+} + O_2^{\overline{\cdot}} . \tag{8}$$

The action of $O_2^{\overline{\cdot}}$ on oxyhemoglobin has been interpreted by a dismutation [237]:

$$HbFe^{3+}—O_2^{\overline{\cdot}} + O_2^{\overline{\cdot}} + 2\,H^+ \rightarrow HbFe^{3+} + O_2 + H_2O_2 . \tag{9}$$

As "oxyhemoglobin" solutions in air are liable to contain appreciable amounts of deoxyhemoglobin, it is astonishing that a reaction of $O_2^{\overline{\cdot}}$ with deoxyhemoglobin has apparently not been considered:

$$HbFe^{2+} + O_2^{\overline{\cdot}} + 2\,H^+ \rightarrow HbFe^{3+} + H_2O_2 . \tag{10}$$

for it would have provided a simple explanation for the observed decrease of methemoglobin formation in the presence of 13.3 mM azide when full oxygenation of hemoglobin was approached [58].

The reaction of $O_2^{\overline{\cdot}}$ with methemoglobin obviously regenerates oxyhemoglobin:

$$HbFe^{3+} + O_2^{\overline{\cdot}} \rightarrow HbFe^{3+}—O_2^{\overline{\cdot}} . \tag{11}$$

The "autoxidation" of bovine heart myoglobin [55] and of shark hemoglobin [53] were followed respectively by the decrease of the absorbance at 581 and at 430 nm. BESOD accelerated definitely the "autoxidation" of the myoglobin but only very slightly that of the hemoglobin. With both proteins there was a partial inhibition by catalase.

As H_2O_2 can be produced by equations (9) and (10) and by the spontaneous or enzymatic dismutation of $O_2^{\bar{\cdot}}$ (Scheme 3), its reaction with deoxyhemoglobin [238] and with methemoglobin [239] must also be considered:

$$HbFe^{2+} + H_2O_2 + 2\,H^+ \rightarrow HbFe(IV) + 2\,H_2O\,. \tag{12}$$

This Fe(IV)-derivative is particularly stable with leghemoglobin, while with other hemoglobins and with myoglobins the oxy form is regenerated with H_2O_2 [238]. The reaction of sufficient concentrations of H_2O_2 with methemoglobin could formally yield the $^{\cdot}OH$ radical:

$$HbFe^{3+} + H_2O_2 + H^+ \rightarrow HbFe(IV) + (^{\cdot}OH) + H_2O\,. \tag{13}$$

Free $^{\cdot}OH$ radicals were, however, not detected, as phenoxy radicals were generated at pH 5 and 7 with human methemoglobin, H_2O_2, and phenols instead of the *o*-hydroxy-cyclohexadienyl radicals observed with Fenton's reagent and phenol [239].

It was shown with photochemically generated $O_2^{\bar{\cdot}}$ that chromatographically purified porcine oxyhemoglobin yielded no methemoglobin. The products originated likely instead from reaction (13) and from subsequent reactions. The process was not inhibited by BESOD, but after the addition of catalase methemoglobin was formed and the reaction was inhibited by BESOD [240], like with erythrocyte lysates [241]. By the way normal human erythrocytes contain about 1 SOD and 1 catalase molecule per 1000 hemoglobin molecules [185].

The importance of reaction (13) was also shown with stimulated human neutrophils, which generate $O_2^{\bar{\cdot}}$ and are thus able to lyse erythrocytes. This hemolysis was inhibited by exogenous (Cu,Zn)-SOD in contrast with the heat-denatured enzyme and with catalase. When hemoglobin in the erythrocytes was converted into methemoglobin by a nitrite treatment, SOD became ineffective, but exogenous catalase protected. The erythrocyte became resistant to hemolysis when treated with carbon monoxide, whereby the formation of methemoglobin was blocked and reaction (13) avoided [242].

The role of $O_2^{\bar{\cdot}}$ and of SOD will further be illustrated by some recent examples.

The binding of α-methyldopa to the membrane of human erythrocytes was increased fifty-fold when (Cu,Zn)-SOD was inhibited by diethyldithiocarbamate. The autoxidation of α-methyldopa was inhibited by BESOD [243].

The insulin producing β cells of the pancreas are selectively destroyed by alloxan, which is reduced there to dialurate. Isolated pancreatic islets of mice were protected against the action of alloxan, by BESOD, by catalase, and by scavengers of $^{\cdot}OH$ radicals [244]. A treatment of rat pancreatic islets with SOD, catalase or diethylenetriaminepenta-acetate attenuated the effect of alloxan, suggesting again a participation of $^{\cdot}OH$ radicals [245].

Cataract was induced in eyes of rabbits by 3-aminotriazole, a catalase inhibitor, whereby the H_2O_2 concentration was raised and (Cu,Zn)-SOD partially inactivated [246].

The stimulation by paraquat of collagen synthesis in organ cultures of neonatal rat lungs was abolished by SOD [247].

To quote an example from a very different field, it was also suggested that sunscald in tomatoes is related to the production of $O_2^{\cdot-}$ and that protection is afforded by (Cu,Zn)-SOD [248].

The role of $O_2^{\cdot-}$ in strand breakage was shown with the DNA of the T7 bacteriophage, in this process ·OH radicals were implicated [249]. The possible relationship with spontaneous cancer has been stressed [250, 251].

SOD and catalase protected the DNA of the double-stranded replicative form of the bacteriophage ΦX174 against inactivation by γ rays [252]. The oxygen effect during X irradiation was shown to be due to $O_2^{\cdot-}$ radicals with *Escherichia coli*. The bacterial cells, grown in a low Fe medium in order to reduce the amount of (Fe)-SOD in the periplasmic space, were protected by BESOD (1.23 μg/ml): the oxygen enhancement ratio fell from 2.4 to 1.4 [253]. When (Cu,Zn)-SOD was inhibited for 95% in Chinese hamster cells by diethyldithiocarbamate, which was washed away as it is a radioprotector, the sensitivity to X irradiation was increased [254].

In order to prolong the plasma half-times (Cu,Zn)-SOD has been attached to dextran and to Ficoll [222], and to polyethylene glycol [222, 255]. The inclusion in liposomes allows, moreover, to direct it selectively to particular organs as a function of the composition of the liposomes [95]. When (Cu,Zn)-SOD from bovine liver was cross-linked to rabbit albumin, the antigenicity in rabbits was decreased [256]. The production of human (Cu,Zn)-SOD by genetic engineering might be worthwhile, although the antigenicity of SOD seems low in mammalian species. It is probably linked with the hypervariable regions: residues 17–30 and 109–111 [104]. As a first step the human placenta proved a useful source for the isolation of the mRNA [257].

Striking clinical results have been reported in inflammatory and auto-immune diseases, although in a somewhat limited number of cases [95, 258]. A low molecular mass chromosome-breaking agent was identified in the serum of patients with systemic *lupus erythematosus*. The chromosome aberrations in cultures of normal lymphocytes in the presence of this factor were reduced to normal values by the addition of BESOD to the culture medium [259].

Intra-articular injections of orgotein proved effective in degenerative joint disease [260]. The side effects of radiation therapy in patients with bladder tumors were also reduced by the administration of SOD [261]. The nephrotoxicity of *cis*-diamine-dichloroplatinum in rats was similarly lowered by the subcutaneous administration of (Cu,Zn)-SOD [262].

4.2 Dismutase Activity of Copper Complexes

The most efficient catalyst for the dismutation of $O_2^{\cdot-}$ is the $[Cu(H_2O)_6]^{2+}$ complex [263] with a second-order rate constant of 8×10^9 M^{-1} s^{-1}, more than twice that of BESOD [116].

A superoxide dismutase activity has been reported for Cu complexes of amino acids and peptides [264–266], salicylates [267], penicillamine [268], and indomethacin [269]. In pulse-radiolysis experiments by the decay of $O_2^{\cdot-}$ at 250 nm a second-order rate constant between 0.8×10^9 and 2.4×10^9 M^{-1} s^{-1} was determined at pH 7.5 for the Cu chelates of salicylates [267] and of 0.4×10^9 at pH 7.0 for the Cu-penicillamine complex [268]. With the xanthine-xanthine oxidase system, however, the activity of the bis-(salicylato)-Cu(II) complex, measured by the reduction of Fe(III)-cytochrome *c*, was about 1000 times smaller per mole of Cu than that of BESOD [270]. The inhibition of the nitroblue tetrazolium reduction by Cu complexes of tyrosine and lysine was also 1000–2000 times less effective per mole of Cu than by BESOD [265]. Could this much lower dismutase activity, observed with the xanthine oxidase system, have been due to a general protein effect, namely a complexation of free copper? But, while in the presence of serumalbumin (0.11 mg/ml) the rate constant with the Cu^{2+} hydrate was reduced from 2.7×10^9 to $0.25 \times 10^9\ M^{-1}\ s^{-1}$, it was only decreased from 0.29×10^9 to $0.15 \times 10^9\ M^{-1}\ s^{-1}$ with the $Cu(Gly\text{-}His)_2$ complex [264].

The lower activity of most chelate complexes than that of the aquo complex of Cu^{2+} was attributed to small amounts of free Cu^{2+} liberated by the labile complexes [271]. The dismutase activity reported for the $[Cu(I)_8Cu(II)_6(D\text{-penicillamine})_{12}Cl]^{5-}$ complex was suppressed by EDTA or by a Chelex 100 treatment [272]. Free Cu^{2+} cations can, however, not explain the very high second-order rate constant of $6 \times 10^9\ M^{-1}\ s^{-1}$ measured at pH 7.0 with the $[Cu_2(\text{indomethacin})_4]$ complex by the decay of $O_2^{\cdot-}$ at 250 nm. With the Cu-histidine complexes, moreover, it was concluded from the influence of pH on the pseudo-first order rate constant of the dismutation of $HO_2^{\cdot}/O_2^{\cdot-}$ that $[CuHis_2H]^{3+}$ was the active species with a second-order rate constant of $3 \times 10^8\ M^{-1}\ s^{-1}$ between pH 2 and 7. It was not possible kinetically to distinguish between the classical and the alternative mechanism, with respectively Cu(I) or a $Cu(II)—O_2^{\cdot-}$ complex as intermediate [273].

A superoxide dismutase activity had been reported for the Fe-EDTA complex in contrast with the inactivity of the Cu-EDTA complex. It was shown, on the contrary, that Fe-EDTA, instead of catalysing the dismutation of $O_2^{\cdot-}$, interferes with the reduction of nitroblue tetrazolium and of Fe(III)-cytochrome *c* in the assays of the dismutase activity [274].

Cu chelates of anti-inflammatory and anti-ulcer drugs were often found more active than the free ligands [275, 276]. The structure of the Cu(II) complexes of the anti-inflammatory drug indomethacin [269] and of the anti-ulcer drug cimetidine [277] was determined by X-ray diffraction.

4.3 Dismutase Activity of Other Copper Enzymes

Human ceruloplasmin inhibits lipid autoxidation induced by ascorbate or inorganic Fe [278]. It is considered an "acute-phase" protein with a beneficial effect in inflammation [279]. It was suggested that ceruloplasmin acts as a scavenger of $O_2^{\cdot-}$ radicals, as it inhibited the reduction of Fe(III)-cytochrome *c* and of nitroblue tetrazolium in the presence of xanthine oxidase, acetaldehyde, and dioxygen as an $O_2^{\cdot-}$-generating system [280]. A mechanism without reduction of Cu^{2+}, similar to that presented for (Cu,Zn)-SOD [163], was also postulated for human ceruloplasmin [281]. No dismutase activity could, however, be detected with $O_2^{\cdot-}$ radicals generated by pulse radiolysis, while type-1 Cu^{2+} was reduced [282]. The inhibition of the Fe-dependent lipid peroxidation by ceruloplasmin was tentatively explained by its ferroxidase activity which lowered the concentration of Fe^{2+} in the system, but ceruloplasmin was also an efficient inhibitor of the Cu^{2+}-catalysed peroxidation [283]. Fe-free transferrin also showed some inhibition [278].

A weak dismutase activity with a second-order rate constant of $3 \times 10^6\ M^{-1}\ s^{-1}$ was estimated for the galactose oxidase of *Polyporus circinatus* from the inhibition at pH 7.8 of the reduction of Fe(III)-cytochrome *c* by $O_2^{\overline{\cdot}}$ radicals generated by the xanthine-xanthine oxidase system [284]. Cytochrome *c* oxidase appeared to have a low superoxide dismutase activity, estimated to 1–3% of that of (Cu,Zn)-SOD, by the inhibition of the reduction of *p*-nitrotetrazolium chloride [285]. Dopamine β-monooxygenase showed a very weak superoxide dismutase activity (10 units/mg against 43,000 units/mg for bovine SOD) [286]. The reduction of nitroblue tetrazolium was, however, found to be inhibited by proteins in general and by heat-inactivated dismutase [287].

In conclusion no other Cu protein has a superoxide dismutase activity comparable to that of (Cu,Zn)-SOD.

5 Radicals as Intermediate

5.1 Superoxide Radicals

Galactose oxidase of *P. circinatus* was apparently inhibited by traces of BESOD. It can be inactivated by H_2O_2 produced in the reaction unless catalase was added. There was an activation by traces of $O_2^{\overline{\cdot}}$. In the absence of oxidants the reaction usually showed an induction period [288]. The enzyme, used at very low concentrations in the assays, was protected by proteins like serumalbumin. SOD did, however, not alter the reaction rate when added after 15 min [289]. This was interpreted by an inactivation of SOD by the H_2O_2 accumulated in the reaction [148], but it could just as well mean that SOD had no effect on the active enzyme, but that it did lower the activation in the induction period. Peroxidase activated galactose oxidase and suppressed the effect of SOD [289]. It did protect the enzyme against H_2O_2 inactivation and could have been responsible for appreciable amounts of $O_2^{\overline{\cdot}}$, produced from O_2 and from radicals formed in its action on a substrate.

There is no valid interpretation for the activation by $O_2^{\overline{\cdot}}$ and by hexacyanoferrate(III), although they fitted nicely in a reaction scheme with Cu(III) as the active species [288]. In the oxidation of an alcohol to an aldehyde Cu(III) would be reduced to Cu(I). In the subsequent reaction of Cu(I) with O_2, Cu(II)$O_2^{\overline{\cdot}}$ was considered an intermediate yielding Cu(III) and H_2O_2. This intermediate would be in a reversible equilibrium with $O_2^{\overline{\cdot}}$ and with the resting Cu(II)-enzyme. This resting enzyme would be oxidized by hexacyanoferrate(III) to the active Cu(III) species. There was unfortunately no indication in X-ray absorption measurements for the formation of Cu(III) with hexacyanoferrate(III) and the resting enzyme [290]. EPR measurements indicated that Cu(II) was present in the active enzyme [14]. It was not possible, moreover, to detect $O_2^{\overline{\cdot}}$ by the reduction of Fe(III)-cytochrome *c* in a galactose oxidase — galactose system [284].

Pig kidney diamine oxidase was inhibited by $CuSO_4$ (for 49% at 0.1 mM) and by the Cu complexes of lysine and tyrosine, but not by (Cu,Zn)-SOD [291]. Similar observations were made with purified bovine plasma amine oxidase [292]. There was only a weak inhibition by bovine (Cu,Zn)-SOD, 0.1 mM Cu(II) (salicylate)$_2$ gave a 40% inhibition, excess lysine suppressed the inhibition by the Cu(II)-lysine complex, perhaps by forming the inactive complex Cu(II) $(lysine)_2$. The inhibition by Cu^{2+} and by Cu(II) $(salicylate)_2$ was essentially non-competitive. These data have been interpreted by a dismutation of $O_2^{\overline{\cdot}}$ by the Cu(II) complexes at the active site, which seems inaccessible to SOD. According to the authors this would implicate $O_2^{\overline{\cdot}}$ as an intermediate.

It has been suggested that the activation of dopamine β-monooxygenase by ascorbate was due to the formation of $O_2^{\overline{\cdot}}$ by reduction of O_2, as the enzyme was activated by $O_2^{\overline{\cdot}}$ and inhibited by BESOD [293]. This inhibition of the enzyme could not be confirmed [294, 295]. Dopamine β-monooxygenase could, moreover, be activated by ascorbate in the absence of O_2 [294]. It was nevertheless activated by reduction with $O_2^{\overline{\cdot}}$ at a rate which was 200 times slower than that with ascorbate. The reduced form of the enzyme was unstable in the absence of substrate, which was interpreted by the mentioned low superoxide dismutase activity [286].

In these enzymatic processes there is no indication for free $O_2^{\overline{\cdot}}$ radicals and no direct proof yet for their presence at the active site.

The reaction with O_2 of the fully reduced bovine heart cytochrome *c* oxidase ($a^{2+}Cu_A^+ \cdot Cu_B^+ a_3^{2+}$) was followed at low temperature by optical and EPR spectroscopy [296]. With the mixed-valence-state cytochrome *c* oxidase ($a^{3+}Cu_A^{2+} \cdot Cu_B^+ a_3^{2+}$) cytochrome a^{3+} and Cu_A^{2+} remained unaffected and the presence of three intermediates could be deduced:

$$[Cu_B\underline{a}_3]^{3+} + O_2 \rightleftharpoons [Cu_B^+\underline{a}_3^{2+\delta} \cdot O_2^{-\delta}] \rightleftharpoons [Cu_B^{2+}\underline{a}_3^{2+} \cdot O_2^-]$$
$$\rightleftharpoons [Cu_B^{2+}\underline{a}_3^{2+} \cdot O_2^{2-} \cdot X^+] \qquad (14)$$

The free radical X^+ could be a protein free radical or a porphyrin π-cation radical [297].

5.2 A Paramagnetic Oxygen Intermediate

On adding dioxygen to the fully reduced laccase of the lacquer tree *Rhus vernicifera*, the type-1 Cu and the type-3 Cu-pair were oxidized in the ms range and an optical intermediate was observed at 360 nm [298]. At liquid helium temperatures an EPR signal was observed, which was tentatively interpreted as due to $O^{\overline{\cdot}}$, as a result of its very short relaxation time and of the increase of its linewidth when the reduced laccase of the fungus *Polyporus versicolor* was treated with $^{17}O_2$ [299]. A similar paramagnetic oxygen intermediate was also observed with the laccase of another lacquer tree *Rhus succedanea* and with ceruloplasmin. The decay of the intermediate at 25 °C ($t_{1/2}$ = 1 s at pH 5.5 with *R. succedanea* laccase) was accompanied by the reoxidation of the type-2 Cu [300]. One would expect, however, such an intermediate to be extremely reactive (See Sect. 3.3), while it was stable in tree laccase depleted of type-2 Cu(II) [301].

A new broad EPR resonance at $g = 5$ was observed by rapid freezing on treating reduced cytochrome *c* oxidase with O_2. It could not be attributed to an oxygen containing radical, as the linewidth was not increased with $^{17}O_2$ [302].

6 Radicals as Reaction Product

In the oxidation of ascorbate by O_2 catalysed by ascorbate oxidase, the formation of the monodehydroascorbate free radical was demonstrated by EPR spectroscopy in a flow cell. A steady state was usually reached within 50 ms. The production of the free radical was also followed by the reduction of Fe(III)-cytochrome *c*. Thus the oxidation of ascorbate occurs in a one-electron step [303]. The formation of the monodehydroascorbate free radical was also measured directly by spectrophotometry at 360 nm, where the free radical shows an absorption maximum [304].

The formation of the monodehydroascorbate free radical by dopamine β-monooxygenase was similarly demonstrated by the acceleration of the reduction of Fe(III)-cytochrome *c* [305] and by direct spectrophotometry at 360 nm [304]. The action of the laccase from the mushroom *Agaricus bisporum* on catechol yielded also the semiquinone as the result of a one-electron transfer [306].

With tyrosinase, on the contrary, a two-electron oxidation occurs, as no EPR signal was detected in the catechol oxidation at pH 5.3 [303]. Melanins are polymerization products of tyrosine, whereby tyrosinase catalyses the first steps: the formation of dopa (3,4-dihydroxyphenylalanine) and of dopaquinone, leading to an indolequinone polymer [307]. The peroxidase mechanism for the conversion of tyrosine into dopa in melanogenesis was not substantiated [308]. In natural and synthetic melanins free radicals of a semiquinone type were detected by EPR: $4\text{–}10 \times 10^{17}$ spins g^{-1} of a hydrated suspension (the material was modified on drying and the number of free spins increased). The fairly symmetrical EPR signal had a *g*-value of 2.004 and a linewidth of 4–10 G [307]. The melanins seem to be natural radical scavengers.

The oxidation of dopamine by ceruloplasmin yielded a free radical with a very short lifetime, characterized by an EPR signal at $g = 2.006$ [309]. Dimethyl-*p*-phenylenediamine was oxidized faster by the enzyme to the free radical Wurster's red, which was further oxidized by the enzyme in another one-electron step. The free radical seemed, moreover, to be stabilized by ceruloplasmin [310].

7 The Action of Radicals on Copper Proteins

The autoxidation of ascorbate, a cosubstrate of dopamine β-monooxygenase, induces the degradation of most proteins including catalase and dopamine β-monooxygenase, but with the exception of (Cu,Zn)-SOD. Catalase protects dopamine β-monooxygenase and is therefore generally added in the assay systems [12]. The apparent activation or rather the stabilization of the enzyme (6.5 μg) by small amounts of catalase (3.1 μg) was enhanced by native but not by boiled SOD (100 μg) and also by similar amounts of serumalbumin (100 μg) or of boiled catalase (65 μg) [295]. The effect of SOD points to the intervention of $O_2^{\cdot-}$ in the autoxidation of ascorbate. Proteins in large amounts could react with the strong oxidizing agent formed in this Udenfriend's system [47], thus protecting the enzyme. The ineffectiveness of boiled SOD could be due to its amino-acid composition (See Sect. 4.1.2). While $O_2^{\cdot-}$ did not inactivate dopamine β-monooxygenase [286], the rate of inactivation in the presence of ascorbate was increased by the addition of Cu^{2+} and was faster than that observed with similar concentrations of Cu^{2+} and H_2O_2 [311]. An inactivation of tyrosinase by ascorbate has also been described [312].

The primary products of the pulse radiolysis [2,313,314] of H_2O by pulses of high-energy electrons (1–15 MeV) are hydrated electrons e^-_{aq} and the free radicals $^{\cdot}OH$ besides smaller amounts of $H^{\cdot}$, H_2O_2, H_2, and H_3O^+.

In the presence of O_2 superoxide radicals are generated:

$$e^-_{aq} + O_2 \rightarrow O_2^{\cdot-} \tag{15}$$

$$H^{\cdot} + O_2 \longrightarrow HO_2^{\cdot} \overset{H_2O}{\rightleftharpoons} H_3O^+ + O_2^{\cdot-} \tag{16}$$

In the presence of N_2O the yield of $^{\cdot}OH$ radicals can be doubled:

$$e^-_{aq} + N_2O + H_2O \rightarrow N_2 + {^{\cdot}OH} + OH^- . \tag{17}$$

The $^{\cdot}OH$ radicals can produce the more selective halide or pseudohalide radicals:

$$^{\cdot}OH + Br^- \xrightarrow{Br^-} Br_2^{\cdot-} + OH^- \tag{18}$$

$$^{\cdot}OH + SCN^- \xrightarrow{SCN^-} (SCN)_2^{\cdot-} + OH^- \tag{19}$$

and the reducing formate radicals from formate:

$$^{\cdot}OH + HCO_2^- \rightarrow CO_2^{\cdot-} + H_2O \tag{20}$$

in the absence of oxygen in order to avoid:

$$CO_2^{\cdot-} + O_2 \rightarrow CO_2 + O_2^{\cdot-} . \tag{21}$$

With carbonate the $^{\cdot}OH$ radicals generate the carbonate radicals:

$$^{\cdot}OH + CO_3^{2-} \rightarrow CO_3^{\cdot-} + OH^- . \tag{22}$$

The 2-methyl-2-hydroxypropylperoxyl radicals were produced from 2-methyl-2-propanol by pulse radiolysis:

$$\begin{matrix} CH_3 \\ | \\ CH_3{-}C{-}OH \\ | \\ CH_3 \end{matrix} + {^{\cdot}OH} \longrightarrow \begin{matrix} {^{\cdot}CH_2} \\ | \\ CH_3{-}C{-}OH \\ | \\ CH_3 \end{matrix} \xrightarrow{O_2} \begin{matrix} CH_2{-}O_2^{\cdot} \\ | \\ CH_3{-}C{-}OH \\ | \\ CH_3 \end{matrix} \tag{23}$$

They were found not to react with BESOD, the rate constant was estimated to be $< 10^8\ M^{-1}\ s^{-1}$, if there was a reaction at all [315]. The reaction of BESOD was also investigated with several other radicals generated by pulse radiolysis. With the semiquinone of riboflavin 5′-phosphate no reaction was detected. The semiquinone of 9,10-anthraquinone-2-sulfonate and the radical anion of 4-nitroacetophenone converted the enzyme into an unreactive form [316].

The inactivation of (Cu,Zn)-SOD from bovine and human erythrocytes by

$Br_2^{\overline{\cdot}}$ radicals, generated by ^{60}Co γ-irradiation in the presence of Br^-, was investigated. The effect due to H˙ atoms was subtracted. The inactivation yields rose sharply above pH 10 [317]. This could be related with a loosening of the active site as a result of the cleavage of histidine-61 from Cu(II) [137]. There was a destruction of histidine and of tyrosine, and unexpectedly of lysine when Cu was present in the active site [318].

In the presence of carbonate the half-life of the enzymatic activity of bovine and human (Cu,Zn)-SOD was considerably reduced on ^{60}Co γ-irradiation, which indicates a more selective reaction of carbonate radicals with the active site [319].

The reduction of laccase by $O_2^{\overline{\cdot}}$ obtained by pulse radiolysis in the presence of O_2 was followed by the decrease of the absorbance at 614 nm of type-1 Cu. Only a very partial reduction was observed (up to 7%). The binding of F^- anions to type-2 Cu lowered the reduction and the reoxidation rates [320].

The reaction of $CO_2^{\overline{\cdot}}$ radicals with human ceruloplasmin produced $RSSR^{\overline{\cdot}}$ radicals. Their first-order decay occurred at the reduction rate of type-1 Cu. The same mechanism was observed with fungal and tree laccase [321].

The active sites of oxyhemocyanins did resist low doses (below 80 krad or 0.8 J g^{-1}) of ^{60}Co γ-rays or of 280 kV X-rays. On subsequent deoxygenation the deoxyhemocyanin (Cu^I Cu^I) of the chelicerate *Limulus polyphemus* was irreversibly transformed into methemocyanin (Cu^{II} Cu^{II}) by the H_2O_2 formed during the irradiation, while the hemocyanin of the gastropod *Busycon canaliculatum* showed a catalase activity [322]. In the absence of oxygen, the deoxyhemocyanins were submitted to up to 8 times higher doses, whereby the active sites were partially inactivated. This action was attributed to ˙OH radicals as it was eliminated by 0.5 M methanol or formate as their scavengers. Cu(II) in the modified active site could be reduced again by irradiation in the presence of 0.5 M formate. The oxygen binding was restored with *B. canaliculatum* in contrast with *L. polyphemus* hemocyanin. There were obviously other modifications in the protein moiety as a result of the irradiation [323].

8 Concluding Remarks

The discovery of the superoxide dismutase activity of erythrocuprein [86] has indeed opened a wide field.

A (Cu,Zn)-SOD, prepared by less drastic methods and showing the highest possible activity, could clarify the kinetic experiments, although the half-site reactivity seems already rather unlikely. The intermediate might be stabilized and characterized in mixed aqueous systems at sub-zero temperatures [324]. A direct comparison under the same experimental conditions of the metal derivatives of SOD of different species could help to clarify their possible differences.

The storage role of (Cu,Zn)-SOD in seeds e.g. [202] seems plausible, when the Cu-carrier function of ceruloplasmin is considered [17]. The lipophilic anti-inflammatory and anti-ulcer Cu-chelates could also raise the Cu concentration in certain tissues and thus enhance their lysyl oxidase activity. But especially Cu(acetylsalicylate)$_2$ inhibited proline,2-oxoglutarate dioxygenase (EC 1.14.11.2) and lysine,2-oxoglutarate dioxygenase (EC 1.14.1.4), which are also important enzymes in the processing of collagen [325].

In the experiments with galactose oxidase and dopamine β-monooxygenase very low enzyme concentrations have often been used. The addition of an inert protein might help to approximate the natural conditions by providing a protection against surface denaturation and against radicals, and by regulating the trace metal concentrations.

The production of human (Cu,Zn)-SOD by genetic engineering could make sufficient amounts available for a further testing of the promising applications in autoimmune and inflammatory diseases, and as an adjuvant in the radio- and chemotherapy of certain types of tumors. In this respect the hemolysis observed as a secondary effect in patients treated with nitrogen mustards could be ascribed to a striking reduction of the (Cu,Zn)-SOD levels in their erythrocytes [326].

References

1. Pryor, W. A.: The role of free radical reactions in biological systems. In: Free Radicals in Biology, Vol. I (Pryor, W. A., ed.), New York-San Francisco-London, Academic Press, 1976, pp. 1–49
2. Willson, R. L.: Free radicals and tissue damage: Mechanistic evidence from radiation studies. In: Biochemical Mechanisms of Liver Injury (Slater, T. F., ed.), London-New York-San Francisco, Academic Press, 1978, pp. 123–224
3. Kägi, J. H. R., Nordberg, M. (eds.): Metallothionein. Proc. First Internat. Meet. Metallothionein and Other Low Molecular Weight Metal-binding Proteins. Zürich, July 17–22, 1978, Basel-Boston-Stuttgart, Birkhäuser, 1979, p. 378
4. Kojima, Y., Kägi, J. H. R.: Metallothionein. Trends Biochem. Sci. *3*, 90 (1978)
5. Malkin, R., Malmström, B. G.: Adv. Enzymol. *33*, 177 (1970)
6. Fee, J. A.: Structure and Bonding, Vol. 23, Berlin-Heidelberg-New York, Springer, 1975, pp. 1–60
7. Gray, H. B.: Electronic absorption spectroscopy. In: Methods for Determining Metal Ion Environments in Proteins: Structure and Function of Metalloproteins (Darnall, D. W., Wilkins, R. G., eds.), New York-Amsterdam-Oxford, Elsevier/North-Holland, 1980, pp. 14–22
8. Cass, A. E. G., Hill, H. A. O.: Copper proteins and copper enzymes. In: Biological Roles of Copper, Ciba Foundation Symposium 79, Amsterdam-Oxford-New York, Excerpta Medica, 1981, in press
9. Boden, N., Holmes, M. C., Knowles, P. F.: Biochem. Biophys. Res. Commun. *57*, 845 (1974)
10. Lontie, R., Gielens, C.: Molluscan and arthropodan haemocyanins. In: Metalloproteins. Structure, Molecular Function and Clinical Aspects (Weser, U., ed.), Stuttgart-New York, Thieme, 1979, pp. 62–72
11. Vanneste, W. H., Zuberbühler, A.: Copper-containing oxygenases. In: Molecular Mechanisms of Oxygen Activation (Hayaishi, O., ed.), New York-London, Academic Press, 1974, pp. 371–404
12. Skotland, T., Ljones, T.: Inorg. Perspect. Biol. Med. *2*, 151 (1979)
13. Blackburn, N. J., Mason, H. S., Knowles, P. F.: Biochem. Biophys. Res. Commun. *95*, 1275 (1980)
14. Reinhammer, B.: The copper-containing oxidases. In: Adv. Inorg. Biochemistry. Vol. 1 (Eichhorn, G. L., Marzilli, L. G., eds.), New York-Amsterdam-Oxford: Elsevier/North-Holland, 1979, pp. 91–118
15. Deinum, J., Reinhammar, B., Marchesini, A.: FEBS Lett. *42*, 241 (1974)
16. Veldsema, A., Van Gelder, B. F.: Biochim. Biophys. Acta *293*, 322 (1973)
17. Frieden, E., Hsieh, H. S.: Adv. Enzymol. *44*, 187 (1976)
18. Rydén, L., Björk, I.: Biochemistry *15*, 3411 (1976)
19. Noyer, M. et al.: Anal. Biochem. *102*, 450 (1980)
20. Azzi, A., Casey, R. P.: Molec. Cell. Biochem. *28*, 169 (1979)
21. Erecińska, M., Wilson, D. F.: Arch. Biochem. Biophys. *188*, 1 (1978)

22. Malmström, B. G.: Biochim. Biophys. Acta *549*, 281 (1979)
23. Ishimura, Y. et al.: J. Biol. Chem. *255*, 3835 (1980)
24. Hirata, F. et al.: Copper content of indoleamine 2,3-dioxygenase. In: Iron and Copper Proteins. Adv. Experim. Med. Biol., Vol. 74 (Yasunobu, K. T., Mower, H. F., Hayaishi, O., eds.), New York-London: Plenum, 1976, pp. 354–357
25. Brady, F. O., Udom, A.: Is indoleamine 2,3-dioxygenase another heme and copper containing enzyme? In: Iron and Copper Proteins. Adv. Experim. Med. Biol., Vol. 74 (Yasunobu, K. T., Mower, H. F., Hayaishi, O., eds.), New York-London: Plenum, 1976, pp. 343–353
26. Taube, H.: J. Gen. Physiol. *49*, Suppl., 29 (1965)
27. Gilbert, D. L.: Anesthesiology *37*, 100 (1972)
28. Samuel, D.: The physical and theoretical chemistry of oxygen. In: Biochemie des Sauerstoffs (Hess, B., Staudinger, Hj., eds.), Berlin-Heidelberg-New York: Springer, 1968, pp. 6–25
29. Hill, H. A. O.: The chemistry of dioxygen and its reduction products. In: Oxygen Free Radicals and Tissue Damage. Ciba Found. Symp. 65. Amsterdam-Oxford-New York: Excerpta Medica, 1979, pp. 5–17
30. Sullivan, J. D., Jr., Ikawa, M.: Biochim. Biophys. Acta *309*, 11 (1973)
31. Bereman, R. D. et al.: Characterization of the copper(II) site in galactose oxidase. In: Bioinorganic Chemistry II. Adv. Chem. Ser. Vol. 162 (Raymond, K. N., ed.), Washington, Amer. Chem. Soc. 1977, pp. 263–280
32. Buffoni, F.: Pharmacol. Res. Commun. *12*, 101 (1980)
33. Farina, B., Faraone Mennella, M. R., Leone, E.: Ital. J. Biochem. *28*, 270 (1979)
34. Shieh, J. J., Yasunobu, K. T.: Purification and properties of lung lysyl oxidase, a copper-enzyme. In: Iron and Copper Proteins, Adv. Experim. Med. Biol. Vol. 74 (Yasunobu, K. T., Mower, H. F., Hayaishi, O., eds.), New York-London, Plenum, 1976, pp. 447–463
35. Lee, M. H., Dawson, C. R.: Meth. Enzymol. *62*, 30 (1979)
36. Knowles, P. F. et al.: Biochem. J. *111*, 53 (1969)
37. Gabig, T. G., Babior, B. M.: J. Biol. Chem. *254*, 9070 (1979)
38. Tauber, A. I., Goetzl, E. J.: Biochemistry *18*, 5576 (1979)
39. Green, T. R., Schaefer, R. E., Makler, M. T.: Biochem. Biophys. Res. Commun. *94*, 262 (1980)
40. Markert, M., Allaz, M.-J., Frei, J.: FEBS Lett. *113*, 225 (1980)
41. Green, M. R. et al.: FEBS Lett. *100*, 23 (1979)
42. Joseph, M. et al.: Clin. Exp. Immunol. *40*, 416 (1980)
43. Holian, A., Daniele, R. P.: FEBS Lett. *108*, 47 (1979)
44. Hirai, K.-I., Ueno, S., Ogawa, K.: Acta Histochem. Cytochem. *13*, 113 (1980)
45. Ueno, S. et al.: Acta Histochem. Cytochem. *12*, 623 (1979)
46. Peskin, A. V., Zbarsky, I. B., Konstantinov, A. A.: FEBS Lett. *117*, 44 (1980)
47. Fee, J. A.: Superoxide, superoxide dismutases, and oxygen toxicity. In: Metal Ion Activation of Dioxygen (Spiro, T. G., ed.), New York-Chichester-Brisbane-Toronto, Wiley-Interscience, 1980, pp. 209–237
48. Bors, W. et al.: Curr. Topics Radiat. Res. Quarterly *9*, 247 (1974)
49. Fridovich, I.: Superoxide dismutase. In: Molecular Mechanisms of Oxygen Activation (Hayaishi, O., ed.), New York-London, Academic Press, 1974, pp. 453–477
50. Misra, H. P.: J. Biol. Chem. *249*, 2151 (1974)
51. Hassan, H. M., Fridovich, I.: Arch. Biochem. Biophys. *196*, 385 (1979)
52. Bus, J. S., Aust, S. D., Gibson, J. E.: Biochem. Biophys. Res. Commun. *58*, 749 (1974)
53. Misra, H. P., Fridovich, I.: J. Biol. Chem. *247*, 6960 (1972)
54. Wever, R., Oudega, B., Van Gelder, B. F.: Biochim. Biophys. Acta *302*, 475 (1973)
55. Gotoh, T., Shikama, K.: J. Biochem. *80*, 397 (1976)
56. Sugawara, Y., Shikama, K.: Eur. J. Biochem. *110*, 241 (1980)
57. Wallace, W. J., Maxwell, J. C., Caughey, W. S.: Biochem. Biophys. Res. Commun. *57*, 1104 (1974)
58. Wallace, W. J., Caughey, W. S.: Superoxide as a participant in the chemistry of oxyhemoglobin. In: Biochemical and Clinical Aspects of Oxygen (Coughey, W. S., ed.), New York-London-Toronto-Sydney-San Francisco, Academic Press, 1979, pp. 69–86
59. Harbour, J. R., Bolton, J. R.: Biochem. Biophys. Res. Commun. *64*, 803 (1975)
60. McCormick, J. P., Thomason, T.: J. Am. Chem. Soc. *100*, 312 (1978)
61. Buettner, G. R., Oberley, L. W.: FEBS Lett. *98*, 18 (1979)

62. Felix, C. C. et al.: Biochem. Biophys. Res. Commun. *84*, 335 (1978)
63. Sawyer, D. T., Gibian, M. J.: Tetrahedron *35*, 1471 (1979)
64. Fee, J. A., Hildenbrand, P. G.: FEBS Lett. *39*, 79 (1974)
65. Wood, P. M.: FEBS Lett. *44*, 22 (1974)
66. Valentine, J. S.: The chemical reactivity of superoxide anion in aprotic versus protic media: A review. In: Biochemical and Clinical Aspects of Oxygen (Caughey, W. S., ed.), New York-London-Toronto-Sydney-San Francisco, Academic Press, 1979, pp. 659–677
67. Cundall, R. B. et al.: Some reactions of oxidizing radicals with enzymes in aqueous solution. In: Radiation Biology and Chemistry. Research Developments (Edwards, H. E. et al., eds.), Amsterdam-Oxford-New York, Elsevier, 1979, pp. 169–178
68. Beauchamp, C., Fridovich, I.: J. Biol. Chem. *245*, 4641 (1970)
69. Bors, W. et al.: Biochem. Biophys. Res. Commun. *70*, 81 (1976)
70. Saran, M. et al.: Int. J. Radiat. Biol. *37*, 521 (1980)
71. McClune, G. J., Fee, J. A.: FEBS Lett. *67*, 294 (1976)
72. Halliwell, B.: FEBS Lett. *72*, 8 (1976)
73. Ferradini, C. et al.: Photochem. Photobiol. *28*, 697 (1978)
74. Walling, C.: Acc. Chem. Res. *8*, 125 (1975)
75. Cohen, G., Sinet, P. M.: Fenton's reagent — Once more revisited. In: Chemical and Biochemical Aspects of Superoxide and Superoxide Dismutase (Bannister, J. V., Hill, H. A. O., eds.), New York-Amsterdam-Oxford, Elsevier/North-Holland, 1980, pp. 27–37
76. McCord, J. M., Day, E. D., Jr.: FEBS Lett. *86*, 139 (1978)
77. Halliwell, B.: FEBS Lett. *92*, 321 (1978)
78. Floyd, R. A., Wiseman, B. B.: Biochim. Biophys. Acta *586*, 196 (1979)
79. Elstner, E. F., Osswald, W., Konze, J. R.: FEBS Lett. *121*, 219 (1980)
80. Sagone, A. L., Jr. et al.: Biochim. Biophys. Acta *628*, 90 (1980)
81. Hoidal, J. R., Beall, G. D., Repine, J. E.: Infect. Immun. *26*, 1088 (1979)
82. Finkelstein, E., Rosen, G. M., Rauckman, E. J.: Arch. Biochem. Biophys. *200*, 1 (1980)
83. Mann, T., Keilin, D.: Nature (Lond.) *142*, 148 (1938)
84. Markowitz, H., Cartwright, G. E., Wintrobe, M. M.: J. Biol. Chem. *234*, 40 (1959)
85. Carrico, R. J., Deutsch, H. F.: J. Biol. Chem. *245*, 723 (1970)
86. McCord, J. M., Fridovich, I.: J. Biol. Chem. *244*, 6049 (1969)
87. Michelson, A. M., McCord, J. M., Fridovich, I.: Superoxide and Superoxide Dismutases, London-New York-San Francisco, Academic Press, 1977, p. 568
88. Bannister, J. V., Hill, H. A. O. (eds.): Chemical and Biochemical Aspects of Superoxide and Superoxide Dismutase, New York-Amsterdam-Oxford, Elsevier/North-Holland, 1980, p. 414
89. Fridovich, I.: Oxygen radicals, hydrogen peroxide, and oxygen toxicity. In: Free Radicals in Biology, Vol. I (Pryor, W. A., ed.), New York-San Francisco-London, Academic Press, 1976, pp. 239–277
90. Fridovich, I.: Chemical aspects of superoxide radical and of superoxide dismutases. In: Biochemical and Medical Aspects of Active Oxygen (Hayaishi, O., Asada, K., eds.), Baltimore-London-Tokyo, University Park Press, 1977, pp. 3–12
91. Fridovich, I.: Science *201*, 875 (1978)
92. Fridovich, I.: Superoxide and superoxide dismutases. In: Advances in Inorganic Biochemistry. Vol. 1 (Eichhorn, G. L., Marzilli, L. G., eds.), New York-Amsterdam-Oxford, Elsevier/North-Holland, 1979, pp. 67–90
93. Fee, J. A.: Structure-function relationships in superoxide dismutases. In: Superoxide and Superoxide Dismutases (Michelson, A. M., McCord, J. M., Fridovich, I., eds.). London-New York- San Francisco, Academic Press, 1977, pp. 173–192
94. Lippard, S. J. et al.: Physical and chemical studies of bovine erythrocyte superoxide dismutase. In: Bioinorganic Chemistry II (Raymond, K. N., ed.), Washington, DC, Adv. Chem. Series 162, Amer. Chem. Soc., 1977, pp. 251–261
95. Michelson, A. M.: Superoxide dismutases. In: Metalloproteins. Structure, Molecular Function and Clinical Aspects (Weser, U., ed.), Stuttgart-New York, Thieme, 1979, pp. 88–116
96. Caughey, W. S.: Biochemical and Clinical Aspects of Oxygen, New York-London-Toronto-Sydney-San Francisco, Academic Press, 1979, p. 866
97. Ciba Foundation Symposium 65: Oxygen Free Radicals and Tissue Damage, Amsterdam-Oxford-New York, Excerpta Medica, 1979, p. 381

98. Hayaishi, O., Asada, K.: Biochemical and Medical Aspects of Active Oxygen, Baltimore-London-Tokyo, University Park Press, 1977, p. 313
99. Johansen, J. T. et al.: Carlsberg Res. Commun. *44*, 201 (1979)
100. Steinman, H. M.: J. Biol. Chem. *255*, 6758 (1980)
101. Steinman, H. M. et al.: J. Biol. Chem. *249*, 7326 (1974)
102. Ammer, D., Lerch, K.: Primary structure of horse liver superoxide dismutase. In: Chemical and Biochemical Aspects of Superoxide and Superoxide Dismutase (Bannister, J. V., Hill, H. A. O., eds.), New York-Amsterdam-Oxford, Elsevier/North-Holland, 1980, pp. 230–236
103. Jabusch, J. R. et al.: Biochemistry *19*, 2310 (1980)
104. Barra, D. et al.: FEBS Lett. *120*, 53 (1980)
105. Calabrese, L. et al.: Eur. J. Biochem. *56*, 305 (1975)
106. Briggs, R. G., Fee, J. A.: Biochim. Biophys. Acta *537*, 100 (1978)
107. Bannister, W. H. et al.: Internat. J. Biochem. *3*, 560 (1972)
108. Richardson, D. C.: The three-dimensional structure of Cu, Zn superoxide dismutase. In: Superoxide and Superoxide Dismutases (Michelson, A. M., McCord, J. M., Fridovich, I., eds.), London-New York-San Francisco, Academic Press, 1977, pp. 217–223
109. Stoesz, J. D., Malinowski, D. P., Redfield, A. G.: Biochemistry *18*, 4669 (1979)
110. McLachlan, A. D.: Nature *285*, 267 (1980)
111. Cass, A. E. G. et al.: Carlsberg Res. Commun. *43*, 439 (1978)
112. Hill, H. A. O. et al.: Biochem. J. *185*, 245 (1980)
113. Burger, A. R. et al.: Biochemistry *19*, 4139 (1980)
114. Cass, A. E. G. et al.: Biochem. J. *183*, 127 (1979)
115. Fielden, E. M. et al.: Biochem. J. *139*, 49 (1974)
116. McAdam, M. E.: Biochem. J. *161*, 697 (1977)
117. Cockle, S. A., Bray, R. C.: Do all the copper atoms in bovine superoxide dismutase function in catalysis? In: Superoxide and Superoxide Dismutases (Michelson, A. M., McCord, J. M., Fridovich, I., eds.), London-New York-San Francisco, Academic Press, 1977, pp. 215–216
118. Zepp, R. A., Chelack, W. S., Petkau, A.: Bovine superoxide dismutase preparations: Comparison of their biochemical and biological characteristics. In: Chemical and Biochemical Aspects of Superoxide and Superoxide Dismutase (Bannister, J. V., Hill, H. A. O., eds.), New York-Amsterdam-Oxford, Elsevier/North-Holland, 1980, pp. 201–211
119. Rigo, A. et al.: Biochem. J. *161*, 27 (1977)
120. Rigo, A. et al.: Biochem. J. *161*, 31 (1977)
121. Bannister, J. V., Anastasi, A., Bannister, W. H.: Biochem. Biophys. Res. Commun. *81*, 469 (1978)
122. Bossa, F. et al.: Studies on the protein structure of copper-zinc superoxide dismutase. In: Chemical and Biochemical Aspects of Superoxide and Superoxide Dismutase (Bannister, J. V., Hill, H. A. O., eds.), New York-Amsterdam-Oxford, Elsevier/North-Holland, 1980, pp. 223–229
123. Marmocchi, F. et al.: FEBS Lett. *94*, 109 (1978)
124. Malinowski, D. P., Fridovich, I.: Biochemistry *18*, 5055 (1979)
125. Weser, U., Hartmann, H. J.: FEBS Lett. *17*, 78 (1971)
126. Beem, K. M., Rich, W. E., Rajagopalan, K. V.: J. Biol. Chem. *249*, 7298 (1974)
127. Fee, J. A.: J. Biol. Chem. *248*, 4229 (1973)
128. Lippard, S. J. et al.: Biochemistry *16*, 1136 (1977)
129. Cass, A. E. G. et al.: Biochem. J. *177*, 477 (1979)
130. Fee, J. A., Briggs, R. G.: Biochim. Biophys. Acta *400*, 439 (1975)
131. Rotilio, G. et al.: J. Biol. Chem. *249*, 3157 (1974)
132. Desideri, A. et al.: FEBS Lett. *89*, 83 (1978)
133. Calabrese, L., Cocco, D., Desideri, A.: FEBS Lett. *106*, 142 (1979)
134. Beem, K. M., Richardson, D. C., Rajagopalan, K. V.: Biochemistry *16*, 1930 (1977)
135. Pantoliano, M. W., McDonnell, P. J., Valentine, J. S.: J. Am. Chem. Soc. *101*, 6454 (1979)
136. Valentine, J. S. et al.: Proc. Natl Acad. Sci. USA *76*, 4245 (1979)
137. Boden, N., Holmes, M. C., Knowles, P. F.: Biochem. J. *177*, 303 (1979)
138. Valentine, J. S., Pantoliano, M. W.: Metal binding properties of copper-zinc superoxide dismutase. In: Chemical and Biochemical Aspects of Superoxide and Superoxide Dismutase

(Bannister, J. V., Hill, H. A. O., eds.), New York-Amsterdam-Oxford, Elsevier/North-Holland, 1980, pp. 254–264
139. Bauer, R. et al.: Biochem. Biophys. Res. Commun. *94*, 1296 (1980)
140. Dunbar, J. C. et al.: Carlsberg Res. Commun. *45*, 349 (1980)
141. Klug-Roth, D., Fridovich, I., Rabani, J.: J. Am. Chem. Soc. *95*, 2786 (1973)
142. Fee, J. A., DiCorleto, P. E.: Biochemistry *12*, 4893 (1973)
143. Hodgson, E. K., Fridovich, I.: Biochem. Biophys. Res. Commun. *54*, 270 (1973)
144. Symonyan, M. A., Nalbandyan, R. M.: FEBS Lett. *28*, 22 (1972)
145. Rotilio, G. et al.: Biochim. Biophys. Acta *302*, 229 (1973)
146. Bray, R. C. et al.: Biochem. J. *139*, 43 (1974)
147. Hodgson, E. K., Fridovich, I.: Biochemistry *14*, 5294 (1975)
148. Hodgson, E. K., Fridovich, I.: Biochemistry *14*, 5299 (1975)
149. Blumberg, W. E. et al.: Biochemistry *17*, 1842 (1978)
150. Sawada, Y., Yamazaki, I.: Biochim. Biophys. Acta *327*, 257 (1973)
151. Rotilio, G., Bray, R. C., Fielden, E. M.: Biochim. Biophys. Acta *268*, 605 (1972)
152. Fee, J. A., Ward, R. L.: Biochem. Biophys. Res. Commun. *71*, 427 (1976)
153. Moss, T. H., Fee, J. A.: Biochem. Biophys. Res. Commun. *66*, 799 (1975)
154. McAdam, M. E. et al.: Biochem. J. *167*, 271 (1977)
155. Bailey, D. B., Ellis, P. D., Fee, J. A.: Biochemistry *19*, 591 (1980)
156. Armitage, I. M. et al.: J. Magn. Reson. *29*, 375 (1978)
157. Malinowski, D. P., Fridovich, I.: Biochemistry *18*, 5909 (1979)
158. Borders, C. L., Jr., Johansen, J. T.: Biochem. Biophys. Res. Commun. *96*, 1071 (1980)
159. Cass, A. E. G., Hill, H. A. O.: Anion binding to copper(I) superoxide dismutase: A high resolution ^{1}H nuclear magnetic resonance spectroscopic study. In: Chemical and Biochemical Aspects of Superoxide and Superoxide Dismutase (Bannister, J. V., Hill, H. A. O., eds.), New York-Amsterdam-Oxford, Elsevier/North-Holland, 1980, pp. 290–298
160. Bielski, B. H. J.: Discussion. Formation of copper-superoxide complexes. In: Ciba Foundation Symp. 65: Oxygen Free Radicals and Tissue Damage, Amsterdam-Oxford-New York, Excerpta Medica, 1979, pp. 52–53
161. Bielski, B. H. J., Chan, P. C.: J. Am. Chem. Soc. *100*, 1920 (1978)
162. Płonka, A., Metodiewa, D., Jeziorowska, A.: Biochem. Biophys. Res. Commun. *81*, 1344 (1978)
163. Płonka, A., Metodiewa, D., Gasyna, Z.: Biochim. Biophys. Acta *612*, 299 (1980)
164. Malinowski, D. P., Fridovich, I.: Biochemistry *18*, 237 (1979)
165. Calabrese, L. et al.: Preparation and preliminary characterization of hybrids of bovine Cu, Zn superoxide dismutase with selectively metal-free and metal-substituted subunits. In: Chemical and Biochemical Aspects of Superoxide and Superoxide Dismutase (Bannister, J. V., Hill, H. A. O., eds.), New York-Amsterdam-Oxford, Elsevier/North-Holland, 1980, pp. 265–271
166. Symonyan, M. A., Nalbandyan, R. M.: Biochem. Biophys. Res. Commun. *90*, 1207 (1979)
167. Weisiger, R. A., Fridovich, I.: J. Biol. Chem. *248*, 4793 (1973)
168. Asada, K., Urano, M., Takahashi, M.-a.: Eur. J. Biochem. *36*, 257 (1973)
169. Salin, M. L., Bridges, S. M.: Z. Pflanzenphysiol. *99*, 37 (1980)
170. Puget, K., Michelson, A. M.: Biochem. Biophys. Res. Commun. *58*, 830 (1974)
171. Vignais, P. M. et al.: Production of superoxide anion and superoxide dismutases in *Paracoccus denitrificans*. In: Chemical and Biochemical Aspects of Superoxide and Superoxide Dismutase (Bannister, J. V., Hill, H. A. O., eds.), New York-Amsterdam-Oxford, Elsevier/North-Holland, 1980, pp. 154–159
172. McCord, J. M.: Iron- and manganese containing superoxide dismutases: Structure, distribution, and evolutionary relationships. In: Iron and Copper Proteins (Yasunobu, K. T., Mower, H. F., Hayaishi, O., eds.), New York-London, Plenum Press, 1976, pp. 540–550
173. Asada, K. et al.: Phylogenic distribution of three types of superoxide dismutase in organisms and in cell organelles. In: Chemical and Biochemical Aspects of Superoxide and Superoxide Dismutase (Bannister, J. V., Hill, H. A. O., eds.), New York-Amsterdam-Oxford, Elsevier/North-Holland, 1980, pp. 136–153
174. McCord, J. M., Keele, B. B., Jr., Fridovich, I.: Proc. Natl Acad. Sci. USA *68*, 1024 (1971)
175. Lynch, R. E., Cole, B. C.: Biochem. Biophys. Res. Commun. *96*, 98 (1980)
176. Forman, H. J., Fridovich, I.: Arch. Biochem. Biophys. *158*, 396 (1973)

177. McCord, J. M., Crapo, J. D., Fridovich, I.: Superoxide dismutase assays: A review of methodology. In: Superoxide and Superoxide Dismutases (Michelson, A. M., McCord, J. M., Fridovich, I., eds.), London-New York-San Francisco, Academic Press, 1977, pp. 11–17
178. Ysebaert-Vanneste, M., Vanneste, W. H.: Analyt. Biochem. *107*, 86 (1980)
179. Misra, H. P., Fridovich, I.: Arch. Biochem. Biophys. *189*, 317 (1978)
180. Hassan, H. M., Dougherty, H., Fridovich, I.: Arch. Biochem. Biophys. *199*, 349 (1980)
181. Peeters-Joris, C., Vandevoorde, A.-M., Baudhuin, P.: Biochem. J. *150*, 31 (1975)
182. Baret, A. et al.: Biochem. Biophys. Res. Commun. *88*, 337 (1979)
183. Rigo, A. et al.: J. Biol. Chem. *254*, 1759 (1979)
184. Sinet, P.-M.: SOD genes in humans: Chromosome localization and electrophoretic variants. In: Superoxide and Superoxide Dismutases (Michelson, A. M., McCord, J. M., Fridovich, I., eds.), London-New York-San Francisco, Academic Press, 1977, pp. 459–465
185. Michelson, A. M. et al.: Clinical aspects of the dosage of erythrocuprein. In: Superoxide and Superoxide Dismutases (Michelson, A. M., McCord, J. M., Fridovich, I., eds.), London-New York-San Francisco, Academic Press, 1977, pp. 467–499
186. Marceau, N., Aspin, N.: Biochim. Biophys. Acta *328*, 351 (1973)
187. Superoxide dismutase as an index of copper, zinc and manganese status, Nutrition Rev. *38*, 326 (1980)
188. Bohnenkamp, W., Weser, U.: 2 Cu, 2 Zn-superoxide dismutase in copper depleted rats. In: Superoxide and Superoxide Dismutases (Michelson, A. M., McCord, J. M., Fridovich, I., eds.), London-New York-San Francisco, Academic Press, 1977, pp. 387–394
189. Alexander, N. M., Benson, G. D.: Life Sciences *16*, 1025 (1975)
190. Burri, B. J. et al.: Clin. Chim. Acta *105*, 249 (1980)
191. Del Villano, B. C. et al.: Science *207*, 991 (1980)
192. Das, S. K., Nair, R. C.: Brit. J. Haematol. *44*, 87 (1980)
193. Loven, D. P. et al.: J. Natl Cancer Instit. *65*, 377 (1980)
194. Reiss, U., Gershon, D.: Eur. J. Biochem. *63*, 617 (1976)
195. Somville, M., Remacle, J.: Arch. Internat. Physiol. Biochim. *88*, B99 (1980)
196. Bannister, W. H., Bannister, J. V.: Biological and Clinical Aspects of Superoxide and Superoxide Dismutase, New York-Amsterdam-Oxford, Elsevier/North-Holland, 1980, p. 434
197. McCord, J. M., Fridovich, I.: Ann. Internal Med. 89, 122 (1978)
198. Oshino, N., Chance, B.: A survey of factors involved in cellular oxygen toxicity. In: Biochemical and Medical Aspects of Active Oxygen (Hayaishi, O., Asada, K., eds.), Baltimore-London-Tokyo, University Park Press, 1977, pp. 191–207
199. Frank, L., Massaro, D.: Am. J. Med. *69*, 117 (1980)
200. Oski, F. A.: New England J. Med. *303*, 454 (1980)
201 Kong, S., Davison, A. J.: Arch. Biochem. Biophys. *204*, 18 (1980)
202. Fee, J. A.: On the question of superoxide toxicity and the biological function of superoxide dismutases. In: Oxidases and Related Redox Systems (King, T. E., Mason, H. S., Morrison, M., eds.), Oxford-New York, Pergamon Press, 1981
203. Hewitt, J., Morris, J. G.: FEBS Lett. *50*, 315 (1975)
204. Hatchikian, C. E., Le Gall, J.: Significance of superoxide dismutase and catalase activities in the strict anaerobes, sulfate reducing bacteria. In: Superoxide and Superoxide Dismutases (Michelson, A. M., McCord, J. M., Fridovich, I., eds.), London-New York-San Francisco, Academic Press, 1977, pp. 159–172
205. Gregory, E. M., Dapper, C. H.: J. Bacteriol. *144*, 967 (1980)
206. Mayeda, E. A., Bard, A. J.: J. Am. Chem. Soc. *96*, 4023 (1974)
207. Finazzi Agró, A. et al.: FEBS Lett. *21*, 183 (1972)
208. Paschen, W., Weser, U.: Biochim. Biophys. Acta *327*, 217 (1973)
209. Weser, U., Paschen, W., Younes, M.: Biochem. Biophys. Res. Commun. *66*, 769 (1975)
210. Poupko, R., Rosenthal, I.: J. Phys. Chem. *77*, 1722 (1973)
211. Nilsson, R., Kearns, D. R.: J. Phys. Chem. *78*, 1681 (1974)
212. Czapski, G., Ilan, Y. A.: Photochem. Photobiol. *28*, 651 (1978)
213. Foote, C. S., Shook, F. C., Abakerli, R. A.: J. Am. Chem. Soc. *102*, 2503 (1980)
214. Goda, K. et al.: Biochem. Biophys. Res. Commun. *58*, 660 (1974)
215. Michelson, A. M.: FEBS Lett. *44*, 97 (1974)
216. Schaap, A. P. et al.: J. Am. Chem. Soc. *96*, 4025 (1974)

217. Hodgson, E. K., Fridovich, I.: Arch. Biochem. Biophys. *172*, 202 (1976)
218. Held, A. M., Hurst, J. K.: Biochem. Biophys. Res. Commun. *81*, 878 (1978)
219. Harrison, J. E., Watson, B. D., Schultz, J.: FEBS Lett. *92*, 327 (1978)
220. Ushijima, Y., Nakano, M.: Biochem. Biophys. Res. Commun. *93*, 1232 (1980)
221. Babior, B. M.: Recent studies on oxygen metabolism in human neutrophils: Superoxide and chemiluminescence. In: Superoxide and Superoxide Dismutases (Michelson, A. M., McCord, J. M., Fridovich, I., eds.), London-New York-San Francisco, Academic Press, 1977, pp. 271–281
222. McCord, J. M., Wong, K.: Phagocyte-produced free radicals: Roles in cytotoxicity and inflammation. In: Oxygen Free Radicals and Tissue Damage, Ciba Foundation Symposium 65, Amsterdam-Oxford-New York, Excerpta Medica, 1979, pp. 343–351
223. Huber, W., Saifer, M. G. P.: Orgotein, the drug version of bovine Cu-Zn superoxide dismutase: I. A summary account of safety and pharmacology in laboratory animals: In: Superoxide and Superoxide Dismutases (Michelson, A. M., McCord, J. M., Fridovich, I., eds.), London-New York-San Francisco, Academic Press, 1977, pp. 517–536
224. Menander-Huber, K. B., Huber, W.: Orgotein, the drug version of bovine Cu-Zn superoxide dismutase. II. A summary account of clinical trials in man and animal. In: Superoxide and Superoxide Dismutases (Michelson, A. M., McCord, J. M., Fridovich, I., eds.), London-New York-San Francisco, Academic Press, 1977, pp. 537–549
225. Petrone, W. F. et al.: Proc. Natl Acad. Sci. USA *77*, 1159 (1980)
226. Ōyanagui, Y.: Biochem. Pharmacol. *25*, 1465 (1976)
227. Bragt, P. C., Bansberg, J. I., Bonta, I. L.: Inflammation *4*, 289 (1980)
228. McCord, J. M.: Science *185*, 529 (1974)
229. Halliwell, B.: FEBS Lett. *96*, 238 (1978)
230. Hofmann, H., Schmut, O.: Albrecht von Graefes Arch. Klin. Ophthalmol. *214*, 181 (1980)
231. Winterbourn, C. C.: Biochem. J. *182*, 625 (1979)
232. Crapo, J. D., Tierney, D. F.: Am. J. Physiol. *226*, 1401 (1974)
233. Hoffman, M., Stevens, J. B., Autor, A. P.: Toxicology *16*, 215 (1980)
234. Liu, J. et al.: J. Appl. Physiol.: Respirat. Environ. Exercise Physiol. *42*, 107 (1977)
235. Bartosz, G., Leyko, W., Fried, R.: Experientia *35*, 5 (1979)
236. Tolmasoff, J. M., Ono, T., Cutler, R. G.: Proc. Natl Acad. Sci. USA *77*, 2777 (1980)
237. Sutton, H. C., Roberts, P. B., Winterbourn, C. C.: Biochem. J. *155*, 503 (1976)
238. Aviram, I., Wittenberg, B. A., Wittenberg, J. B.: J. Biol. Chem. *253*, 5685 (1978)
239. Shiga, T., Imaizumi, K.: Arch. Biochem. Biophys. *154*, 540 (1973)
240. Lynch, R. E. et al.: Biochemistry *16*, 4563 (1977)
241. Lynch, R. E., Lee, G. R., Cartwright, G. E.: J. Biol. Chem. *251*, 1015 (1976)
242. Weiss, S. J.: J. Biol. Chem. *255*, 9912 (1980)
243. Green, F. A., Jung, C. Y., Hui, H.: Biochem. Biophys. Res. Commun. *95*, 1037 (1980)
244. Grankvist, K. et al.: Biochem. J. *182*, 17 (1979)
245. Fischer, L. J., Hamburger, S. A.: Diabetes *29*, 213 (1980)
246. Bhuyan, K. C., Bhuyan, D. K.: Biochim. Biophys. Acta *542*, 28 (1978)
247. Hussain, M. Z., Bhatnagar, R. S.: Biochem. Biophys. Res. Commun. *89*, 71 (1979)
248. Rabinowitch, H. D., Sklan, D.: Planta *148*, 162 (1980)
249. Lesko, S. A., Lorentzen, R. J., Ts'o, P. O. P.: Biochemistry *19*, 3023 (1980)
250. Oberley, L. W., Oberley, T. D., Buettner, G. R.: Med. Hypotheses *6*, 249 (1980)
251. Totter, J. R.: Proc. Natl Acad. Sci. USA *77*, 1763 (1980)
252. Van Hemmen, J. J., Meuling, W. J. A.: Biochim. Biophys. Acta *402*, 133 (1975)
253. Oberley, L. W. et al.: Radiat. Res. *68*, 320 (1976)
254. Westman, G., Marklund, S. L.: Radiat. Res. *83*, 303 (1980)
255. Pyatak, P. S., Abuchowski, A., Davis, F. F.: Res. Commun. Chem. Pathol. Pharmacol. *29*, 113 (1980)
256. Wong, K., Cleland, L. G., Poznansky, M. J.: Agents and Actions *10*, 231 (1980)
257. Bannister, J. V., Hill, H. A. O., Bannister, W. H.: FEBS Lett. *121*, 215 (1980)
258. Michelson, A. M., Puget, K.: Compt. Rend. Soc. Biol. *173*, 380 (1979)
259. Emerit, I. et al.: Hum. Genet. *55*, 341 (1980)
260. Lund-Olesen, K., Menander, K. B.: Curr. Therap. Res. *16*, 706 (1974)
261. Menander-Huber, K. B., Edsmyr, F., Huber, W.: Urol. Res. *6*, 255 (1978)
262. McGinness, J. E. et al.: Physiol. Chem. Physics *10*, 267 (1978)

263. Rabani, J., Klug-Roth, D., Lilie, J.: J. Phys. Chem. *77*, 1169 (1973)
264. Brigelius, R. et al.: FEBS Lett. *47*, 72 (1974)
265. Younes, M., Weser, U.: FEBS Lett. *61*, 209 (1976)
266. Klug-Roth, D., Rabani, J.: J. Phys. Chem. *80*, 588 (1976)
267. Younes, M. et al.: Biochem. Biophys. Res. Commun. *81*, 576 (1978)
268. Lengfelder, E. et al.: Biochim. Biophys. Acta *567*, 492 (1979).
269. Weser, U. et al.: Biochim. Biophys. Acta *631*, 232 (1980)
270. deAlvare, L. R., Goda, K., Kimura, T.: Biochem. Biophys. Res. Commun. *69*, 687 (1976)
271. Pasternack, R. F., Skowronek, W. R., Jr.: J. Inorg. Biochem. *11*, 261 (1979)
272. Robertson, P., Jr., Fridovich, I.: Arch. Biochem. Biophys. *203*, 830 (1980)
273. Weinstein, J., Bielski, B. H. J.: J. Am. Chem. Soc. *102*, 4916 (1980)
274. Diguiseppi, J., Fridovich, I.: Arch. Biochem. Biophys. *203*, 145 (1980)
275. Sorenson, J. R. J.: J. Med. Chem. *19*, 135 (1976)
276. Brown, D. H. et al.: J. Med. Chem. *23*, 729 (1980)
277. Greenaway, F. T. et al.: J. Am. Chem. Soc. *102*, 7782 (1980)
278. Al-Timimi, D. J., Dormandy, T. L.: Biochem. J. *168*, 283 (1977)
279. Denko, C. W.: Agents and Actions *9*, 333 (1979)
280. Goldstein, I. M. et al.: J. Biol. Chem. *254*, 4040 (1979)
281. Płonka, A. et al.: Biochem. Biophys. Res. Commun. *95*, 978 (1980)
282. Bannister, J. V. et al.: FEBS Lett. *118*, 127 (1980)
283. Gutteridge, J. M. C., Richmond, R., Halliwell, B.: FEBS Lett. *112*, 269 (1980)
284. Cleveland, L., Davis, L.: Biochim. Biophys. Acta *341*, 517 (1974)
285. Markossian, K. A. et al.: Biochem. Biophys. Res. Commun. *81*, 1336 (1978)
286. Henry, J.-P., Hirata, F., Hayaishi, O.: Biochem. Biophys. Res. Commun. *81*, 1091 (1978)
287. Amano, D. et al.: Biochem. Biophys. Res. Commun. *66*, 272 (1975)
288. Hamilton, G. A. et al.: J. Am. Chem. Soc. *100*, 1899 (1978)
289. Kwiatkowski, L. D., Kosman, D. J.: Biochem. Biophys. Res. Commun. *53*, 715 (1973)
290. Blumberg, W. E. et al.: Is trivalent copper a viable oxidation state in the enzymic turnover of copper proteins? In: Oxidases and Related Redox Systems (King, T. E., Mason, H. S., Morrison, M., eds.), Oxford-New York, Pergamon Press, 1981
291. Younes, M., Weser, U.: Biochim. Biophys. Acta *526*, 644 (1978)
292. Dooley, D. M., Coolbaugh, T. S.: Biochem. Biophys. Res. Commun. *96*, 823 (1980)
293. Liu, T. Z., Shen, J.-T., Ganong, W. F.: Proc. Soc. Exp. Biol. Med. *146*, 37 (1974)
294. Diliberto, E. J., Jr., Kaufman, S.: Lack of inhibition of dopamine-β-hydroxylase by superoxide dismutase. In: Superoxide and Superoxide Dismutases (Michelson, A. M., McCord, J. M., Fridovich, I., eds.), London-New York-San Francisco, Academic Press, 1977, pp. 407–408
295. Brown, F. C., Harralson, J. D., Zawad, J. S.: Neurochem. Res. *3*, 357 (1978)
296. Clore, G. M. et al.: Biochem. J. *185*, 139 (1980)
297. Clore, G. M. et al.: Biochem. J. *185*, 155 (1980)
298. Andréasson, L.-E., Brändén, R., Reinhammar, B.: Biochim. Biophys. Acta *438*, 370 (1976)
299. Aasa, R. et al.: Biochem. Biophys. Res. Commun. *70*, 1204 (1976)
300. Brändén, R., Deinum, J.: Biochim. Biophys. Acta *524*, 297 (1978)
301. Malmström, B. G.: The mechanism of dioxygen reduction in cytochrome *c* oxidase and laccase. In: Oxidases and Related Redox Systems (King, T. E., Mason, H. S., Morrison, M., eds.), Oxford-New York, Pergamon Press, 1981
302. Shaw, R. W., Hansen, R. E., Beinert, H.: J. Biol. Chem. *253*, 6637 (1978)
303. Yamazaki, I.: Free radicals in enzyme-substrate reactions. In: Free Radicals in Biology. Vol. II (Pryor, W. A., ed.), New York-San Francisco-London, Academic Press, 1977, pp. 183–218
304. Skotland, T., Ljones, T.: Biochim. Biophys. Acta *630*, 30 (1980)
305. Ljones, T., Skotland, T.: FEBS Lett. *108*, 25 (1979)
306. Nakamura, T., Ogura, Y.: Electron spin resonance studies on mushroom laccase. In: Magnetic Resonance in Biological Systems (Ehrenberg, A., Malmström, B. G., Vänngård, T., eds.), Oxford-London-Edinburgh-New York-Toronto-Sydney-Paris-Braunschweig, Pergamon Press, 1967, pp. 205–212
307. Sealy, R. C. et al.: Structure and reactivity of melanins: Influence of free radicals and metal ions.

In: Free Radicals in Biology, Vol. IV (Pryor, W. A., ed.), New York-San Francisco-London, Academic Press, 1980, pp. 209–259
308. Smith, P. I., Swan, G. A.: Biochem. J. *153*, 403 (1976)
309. Walaas, E., Walaas, O., Løvstad, R.: The interaction of ceruloplasmin with catecholamines. In: The Biochemistry of Copper (Peisach, J., Aisen, P., Blumberg, W. E., eds.), New York-London, Academic Press, 1966, pp. 537–544
310. Walaas, E., Løvstad, R. A., Walaas, O.: Arch. Biochem. Biophys. *121*, 480 (1967)
311. Skotland, T., Ljones, T.: Arch. Biochem. Biophys. *201*, 81 (1980)
312. Baruah, P., Swain, T.: Biochem. J. *55*, 392 (1953)
313. Adams, G. E., Wardman, P.: Free radicals in biology: The pulse radiolysis approach. In: Free Radicals in Biology, Volume III (Pryor, W. A., ed.), New York-San Francisco-London, Academic Press, 1977, pp. 53–95
314. Klapper, M. H., Faraggi, M.: Quart. Rev. Biophys. *12*, 465 (1979)
315. O'Neill, P., Fielden, E. M.: Pulse radiolysis investigation of the interaction of bovine superoxide dismutase with organic free radicals. In: Chemical and Biochemical Aspects of Superoxide and Superoxide Dismutase (Bannister, J. V., Hill, H. A. O., eds.), New York-Amsterdam-Oxford, Elsevier/North-Holland, 1980, pp. 357–363
316. Wardman, P.: Specificity of superoxide dismutase in catalysing redox reactions: A pulse radiolysis study. In: Radiation Biology and Chemistry. Research Developments (Edwards, H. E. et al., eds.), Amsterdam-Oxford-New York, Elsevier, 1979, pp. 189–196
317. Roberts, P. B. et al.: Radiat. Res. *60*, 441 (1974)
318. Barra, D. et al.: Biochem. Biophys. Res. Commun. *64*, 1303 (1975)
319. Maral, J., Michelson, A. M., Monny, C.: Quoted in the discussion. In: Oxygen Free Radicals and Tissue Damage, Ciba Foundation Symp. 65, Amsterdam-Oxford-New York, Excerpta Medica, 1979, pp. 36–37
320. Pecht, I., Farver, O., Goldberg, M.: Electron transfer pathways in blue copper proteins. In: Bioinorganic Chemistry II. Advances in Chemistry Series, Vol. 162 (Raymond, K. N., ed.), Washington, Amer. Chem. Soc., 1977, pp. 179–206
321. Pecht, I., Goldberg, M.: Electron transfer pathways to and within redox proteins: Pulse radiolysis studies. In: Fast Processes in Radiation Chemistry and Biology (Adams, G. E., Fielden, E. M., Michael, B. D., eds.), Wiley, Chichester-New York-Toronto-Sydney, 1975, pp. 277–284
322. Schubert, J., White, E. R., Becker, L. F., Jr.: Radiation chemical studies on oxygen-carrying proteins: Hemocyanin. In: Radiation Chemistry I. Adv. Chem. Ser. Vol. 81 (Gould, R. F., ed.), Washington, Amer. Chem. Soc., 1968, pp. 480–495
323. Ke, C. H., Schubert, J.: Radiat. Res. *49*, 507 (1972)
324. Douzou, P.: Cryobiochemistry, An Introduction, London-New York-San Francisco, Academic Press, 1977, p. 286
325. Myllylä, R. et al.: Biochem. Biophys. Res. Commun. *89*, 98 (1979)
326. Auclair, C., Dhermy, D., Boivin, P.: Cancer Chemother. Pharmacol. *4*, 281 (1980)

Note Added in the Proof

2 Copper Proteins in General

Although the stoichiometry of the copper atoms in ceruloplasmin is not settled yet, two type-1 copper atoms are now generally accepted, besides one type-2 copper and one type-3 copper pair, and possibly a sixth and a loose seventh copper atom.

4.1.3 Quaternary Structure and Half-Site Reactivity

The dissociation of SOD observed by gel chromatography in 8 M urea, in contrast with the sedimentation equilibrium data, might be due to differences in the experimental conditions such as concentration effects.

The Metabolism of Superoxide Anion and Its Progeny in Blood Cells

Robert E. Lynch

University of Utah College of Medicine, Salt Lake City, Utah 84132*, USA

Table of Contents

* Supported by Grant # AM 20207 from the National Institutes of Health

Abbreviations

PMNs — polymorphonuclear neutrophils
CGD — chronic granulomatous disease
FMLP — formyl methionyl leucyl phenylalanine
FMP — formyl methionyl phenylalanine
FMV — formyl methionyl valine
FMA — formyl methionyl alanine
FnorLLP — formyl nor leucyl leucyl phenylalanine
FMMM — formyl $(\text{methionine})_3$
FMLL — formyl methionyl $(\text{leucine})_2$
FML — formyl methionyl leucine
FMLGlu — formyl methionyl leucyl glutamic acid
MMMM — $(\text{methionine})_4$
KMB — 2-keto-4-thio methylbutyric acid

1 Introduction

The study of cells in blood has often provided the first and most detailed information on the biochemistry, physiology, immunology, genetics and molecular biology of processes which occur in eukaryotic cells. Much of the credit for the pioneering work on blood cells must go to the ease with which these cells are obtained in homogeneous suspension and to the simplicity of returning them to their sites of action through intravenous infusion. In keeping with this precedent blood cells have occupied a central role in the investigation of the metabolism of radicals derived from molecular oxygen.

The 'modern' era of the study of the species derived from the reduction of molecular dioxygen began with the discovery by McCord and Fridovich of the enzymic function of erythrocuprein [1], a cupro-enzyme first isolated from erythrocytes. The reaction catalyzed by erythrocuprein in which superoxide anions dismute,

$$O_2^- + O_2^- + 2\,H^+ \rightarrow H_2O_2 + O_2\,, \tag{1}$$

caused it to be renamed superoxide dismutase. This momentous discovery implied that at least one free radical occurred in cells in sufficient concentrations that an enzyme had evolved which catalyzed its destruction. This enzyme could be shown to be present in cells which were obliged to survive in the presence of molecular oxygen. Since free radicals were nearly all reactive and unstable, it was proposed that superoxide anion, O_2^-, was an agent of injury of the cells in which it was produced. Superoxide dismutase, then, acted to forbid the injury which could be inflicted on cells by their inadvertant reduction of dioxygen by one electron to form O_2^-, rather than by four electrons to form water. A variety of data which support this proposal have been summarized [2].

Just as the erythrocyte provided the first homogeneous superoxide dismutase the granulocyte has been the exemplar of the cell which forms O_2^- in a controlled and apparently purposive manner. In the years following the demonstration by Babior et al. that polymorphonuclear neutrophils (PMNs) formed O_2^- when stimulated with latex particles[3] an extensive literature has appeared devoted to the means by which these cells are made to form O_2^-. In this review I shall recount the results of investigations of the role of O_2^- and its progeny in the metabolism of both PMNs and red blood cells.

2 Polymorphonuclear Leukocytes

These respiratory burst in PMNs refers to the dramatic, but self-limited, increase in the consumption of oxygen exhibited by PMNs which are stimulated with a variety of agents. Although known since 1933[4] the understanding of the respiratory burst was advanced by the demonstration that one product of the reduction of dioxygen by PMNs was H_2O_2[5]. The involvement of the respiratory burst in the microbicidal function of PMNs was suspected in the 1960's when it was demonstrated that killing of certain microorganisms by PMNs was reduced under hypoxic conditions[6] and that in chronic granulomatous disease (CGD), defective killing of ingested bacteria was associated with an inability to undergo the respiratory burst[7,8]. It was proposed that products of the burst were bactericidal.

The demonstration that PMNs formed O_2^- in the respiratory burst[3] necessitated the consideration of all the species which result when dioxygen is reduced one electron at a time (Fig. 1). Superoxide, the result of the reduction of dioxygen by one electron, appears to act mainly as a mild reductant in aqueous solutions. But when it coexists with H_2O_2, its spontaneous dismutation product, O_2^- can initiate a number of potentially injurious events [reviewed by Fridovich[2]]. The primary means by which cells deal with superoxide anions appears to be through the catalysis of their dismutation by a family of metalloenzymes collectively designated superoxide dismutases.

Reduction of O_2^- by another electron (Fig. 1) results in the formation of H_2O_2 which cells can eliminate through a reaction which is catalyzed by catalase and is analogous to the one catalyzed by the superoxide dismutases. Peroxidases also catalyze the destruction of H_2O_2 using an organic molecule such as glutathione as the reducing substrate. Hydrogen peroxide acts mainly as a mild oxidant but with the

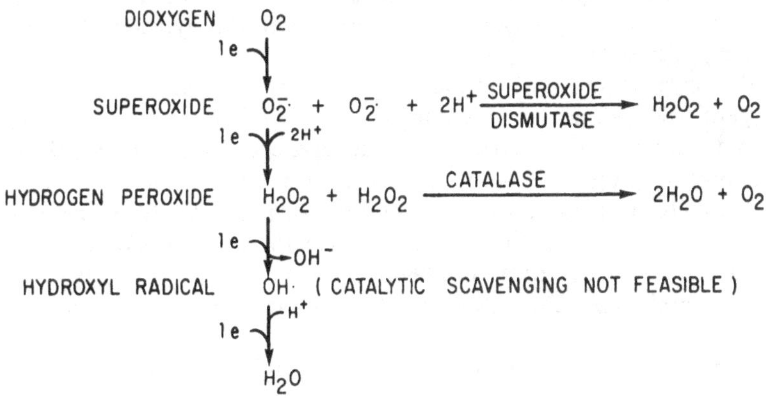

Fig. 1. The steps in the reduction of dioxygen to water

capacity, in concentrations great enough, to induce severe cytocidal effects. Probably more germane to the reactions in which it engages in PMNs is the ability of H_2O_2 to act as a substrate for myeloperoxidase in which capacity it engenders a very potent oxidant(s). The reaction(s) catalyzed by myeloperoxidase uses H_2O_2 and a halide as substrate to form a reactive product believed to be hypochlorous acid:

$$H_2O_2 + Cl^- \rightarrow HOCl + OH^- . \quad (2)$$

Further reaction of HOCl with H_2O_2 can yield singlet oxygen:

$$OCl^- + H_2O_2 \rightarrow Cl^- + H_2O + {}^1O_2 . \quad (3)$$

A number of the reactions in which HOCl and ${}^1O_2$ might participate to kill bacteria and to attack biological molecules have been documented. These include the halogenation of tyrosines, the formation of aldehydes and chloramines, the attack of ${}^1O_2$ on unsaturated bonds in fatty acids, and decarboxylation of amino acids. The experimental basis for these reactions has been reviewed by Klebanoff and Clark [9].

Further reduction of H_2O_2 yields the hydroxyl radical, $OH^{\cdot}$, indeed a powerful but indiscriminant and very unstable oxidant. Superoxide anion may itself serve as the reductant of H_2O_2 to form $OH^{\cdot}$ in the presence of ionic iron in a reaction called the Haber-Weiss reaction [10]:

$$O_2^- + H_2O_2 \xrightarrow{Fe^{3++}} OH^{\cdot} + OH^- + O_2 . \quad (4)$$

Ferric iron appears to undergo cyclic reduction by O_2^- to Fe^{2+} and oxidation of Fe^{2+} by H_2O_2 [11,12].

$$O_2^- + Fe^{3+} \rightarrow Fe^{2+} + O_2 \quad (5)$$

$$Fe^{2+} + H_2O_2 \rightarrow OH^{\cdot} + OH^- + Fe^{3+} \quad \text{[The Fenton}^{13)}\text{ reaction]} . \quad (6)$$

Finally, when $OH^{\cdot}$ is reduced by one electron it forms OH^-. The extreme reactivity of $OH^{\cdot}$ may preclude the existence of an enzyme which could catalytically scavenge it; in any event, none is known.

Singlet molecular oxygen, ${}^1O_2$, is also thought to arise, perhaps in reaction 4 or 5 (in place of dioxygen). The addition of ${}^1O_2$ completes the cast of characters comprising molecules which may mediate the effects of O_2^-, since whenever O_2^- is formed, H_2O_2, $OH^{\cdot}$ and ${}^1O_2$ all may exist in aqueous solutions in the presence of micromolar concentrations of ionic iron which contaminate many buffers. Armed with the capacity to form species with such a variety of affinities for electrons, the PMN is endowed with the ability to initiate a formidable array of chemical reactions, not merely between the various species themselves but also with its own constitutents and those of ingested microbes.

That O_2^- was the product of the burst was first shown by Babior [3] who demonstrated in suspensions of intact PMNs to which latex particles were added as a phagocytosable stimulus that cytochrome c was reduced, a characteristic reaction of O_2^-. The addition of superoxide dismutase abolished the reduction of cytochrome c, demonstrating by the specificity of the reaction which superoxide dismutase catalyzes, that the reductant was indeed O_2^- [3]. In experiments of this type in which the enzymic action of superoxide dismutase is used to show that O_2^- is a necessary reactant, controls are frequently included which demonstrate that the peptide of superoxide dismutase which has lost catalytic activity (due to heating or demetallation) has also lost its effect since reactions have been described in which the catalytically inactive enzyme acts as an inhibitor, by some mechanism independent of its dismutation of O_2^- [14]. Further evidence that the effect of superoxide dismutase is an action on O_2^- is provided by demonstrating that the effect is seen at such low concentrations of the enzyme that a catalytic effect is highly likely and by reproducing the effect with an unrelated superoxide dismutase such as the enzyme containing Fe from *E. coli.*

The elicitation of the respiratory burst has proven to be a complex phenomenon. The central event is the activation of the enzyme which reduces O_2 to form O_2^-, an enzyme usually termed "the oxidase." A minimum number of steps resulting in the explosive but self-limited production of O_2^- include stimulation of the cell by some interaction between an external agent and the cell membrane, transmission of the signal from the surface of the cell to the apparatus which activates the oxidase, activation of the oxidase, catalytic oxidation of substrates yielding O_2^- and finally, inactivation of the oxidase. By a continuous recording of the formation of O_2^- spectrophotometrically it has been possible to subdivide the burst into separate phases [15]. A suspension of human PMNs in the sample beam of the spectrophotometer was stimulated to form O_2^- in the presence of cytochrome c. The change in absorbance was recorded continuously against a cuvette with identical components except that superoxide dismutase was present to scavenge O_2^-. The change in absorbance represented the O_2^--dependent reduction of cytochrome c. After the addition of phorbol myristate acetate there was a lag during which no formation of O_2^- occurred near the end of which the formation of O_2^- commenced, gradually accelerating to a limiting linear rate. The rate of formation of O_2^- subsequently declined, corresponding to the termination of the respiratory burst. The duration of the lag was designated the period of activation and the linear rate of reduction of cytochrome c was termed the activity. The lag time and the activity were found in some cases to vary independently. As the temperature was raised from 26° to 46° the lag time shortened progressively whereas the activity was maximal at 37° to 40°, declining at greater temperatures. Activation was relatively independent of pH but the activity exhibited a sharp optimum at pH 7.5. Calcium ions and magnesium ions affected neither the lag nor the activity when PMA was the stimulus but when opsonized zymonsan was the stimulus both ions diminished the lag and increased the activity. N-ethyl maleimide had little or no effect on the lag time but progressively reduced the activity. The activity was markedly diminished by 1 mM 2-deoxyglucose and was diminished further by the addition of both 2-deoxyglucose and 0.5 mM NaCN, whereas CN^- alone had no effect. Profound depletion of ATP by the combination of 2-deoxyglucose and CN^- nearly abolished the activity without affecting the lag time. By this subdivision of the respiratory burst the lag time encompasses the interaction of stimulus with receptor, the transmission of the signal, and the activation. The activity is the phase of turnover of the oxidase. The phase of declining activity represents inactivation. The stimulation of the burst will therefore be discussed as if these were separable components for the sake of simplicity although they are clearly intimately interrelated.

2.1 Stimuli for the Respiratory Burst

A very wide variety of materials elicit the burst including detergents, opsonized microorganisms, aggregated immunoglobulins, fatty acids, lectins, fungal products, Ca ionophores, chemotactic peptides, polysaccharides, inorganic anions and plastics. The bewildering array of active agents suggests that there must be several consecutive sites at which the activation of the respiratory burst can be initiated simply because it seems so unlikely that any single target could be responsive in the same way to such diverse materials.

Of the particulate stimuli certain ones are far more active when they are coated with proteins from serum (opsonized) than when they are not. However, others like latex beads elicit formation of O_2^- by PMNs without opsonization [3]. DeChatelet et al. [16] found that the production of O_2^- by PMNs from man and rabbit was stimulated by opsonized but not unopsonized zymosan (fragments of yeast cell walls). Bacteria alone were found to stimulate the formation of O_2^- but in the presence of serum bacteria stimulated the formation of O_2^- three fold [17]. However, the stimulatory effect of bacteria appeared to be caused by changes which the bacteria produced by an interaction with constitutents of serum, because serum itself after exposure to the bacteria stimulated production of O_2^- by PMNs. The active component from serum was heat sensitive (100°) and not sedimentable at 105,000 g. Whether this material was derived from the components of serum or from the bacteria is not clear but may have been a protein of the complement system.

Immunoglubulins and components of complement stimulate formation of O_2^-. The proteolytic fragment of C_5, C_{5a}, prepared by gel filtration of serum treated with zymosan, and IgG aggregated by heat both elicited formation of O_2^- [18]. The ability of IgG to elicit the formation of O_2^- by PMN's was also demonstrated when PMNs were exposed to IgG and IgA bound to polystyrene plates [19] or to IgG which was heat-aggregated and bound to micropore filters [20].

Both IgG and C_{3b} were found to stimulate the respiratory burst separately and independently [21] in one study but in another [22] C_{3b} was active only in the presence of IgG. The divergent findings may be attributable to differing methods. In one study [21] serum was mixed with agarose beads which fix complement with the binding of C_{3b}. One set of beads was then heated at 50° for 30 min to remove C_{3b} and the other was boiled in 2 M NaCl to remove IgG. The completeness of the removal of each component was verified by the loss of reactivity of the beads with a fluorescent monospecific antibody. Beads prepared in this way with either IgG or with C_{3b} attached elicited the formation of O_2^- by PMNs whereas the release of lysosomal contents occurred only with beads containing both IgG and C_{3b}. The effects of the IgG-agarose were blocked by $F(ab')_2$ and those of C_{3b} were blocked by antiserum to C_3.

In the other experiments [22] C_3 was purified chromatographically and C_{3b} was generated from the C_3 by digestion with trypsin. The C_{3b} was allowed to attach to red cells which could then bind to PMNs. The specificity of the binding of the C_{3b}-red cells to PMNs was verified by showing inhibition by C_{3b} and by C_{3b}-inactivator in the fluid phase. The C_{3b}-coated red cells did not elicit the formation of O_2^- whereas IgG attached to red cells did. However, the red cells with C_{3b} attached did augment the stimulation by IgG-coated red cells of production of O_2^- and of the release of lysosomal enzymes.

Clearly a number of differences could explain these divergent results. Boiling could cause loss of antigenicity but retention of activity in stimulating the formation of O_2^-. Alternately, the C_{3b} might be attached to agarose beads and to erythrocytes in different ways so that different parts of the molecule were exposed to the PMNs. The assays used to detect contamination of C_{3b} by IgG may also differ in sensitivity (immunofluorescence vs. gel electrophoresis of isolated C_3).

The chemotactic peptides are also active in the stimulation of release of O_2^- and in the release of lysosomal contents [23-25]. Directed locomotion by PMNs is stimulated by a family of tri-, tetra-, and dipeptides which have in common formylated methionine as the first amino acid. Since formylated methionine is the initiator to which other amino acids are added in the synthesis of bacterial protein, small peptides which commence with formylated methionine are likely to be liberated proteolytically at sites of bacterial infection, possibly explaining their great potency as chematoxins. Using superoxide-dependent chemiluminescence (see below) as a measure of synthesis of O_2^-, Hatch et al. [23] showed a hierarchy of potency of FMLP > FMP > FMV > FMA. Becker et al. [24] measured the formation of O_2^- in suspensions of rabbit PMNs and found that the hierarchy of potency was the same for stimulation of the formation of O_2^- as it was for stimulation of release of lysosomal enzymes, namely

FMLP > FnorLLP > FMMM > FMLL > FML
> FMLGlu > MMMM .

In all cases the concentration of the peptide which caused $^1/_2$ maximal stimulation of either chemotaxis or secretion of lysosomal enzymes was less than the concentration required to produce $^1/_2$ maximal stimulation of formation of O_2^-. Specificity was shown by the competitive antagonism by the analogue of FMLP, carbobenzoxy-phenyl alanyl-methionine. Like Becker et al., Lehmyer et al. [25] found that the concentration of FMLP required to induce the respiratory burst was greater than that needed to stimulate chemotaxis. They speculated that the high concentrations which one might expect to find at the origin of a stimulus of chemotaxis, such as a bacterial infection, might provoke the PMN which arrived at the source of the signal to liberate the reactive metabolites of molecular oxygen. The hierarchy of response to chemotactic peptides of PMNs was not the same as in monocytes [26] nor was the interaction between separate stimuli. The chemotactic peptides can also enhance the stimulation elicited by other agents; exposure of PMNs to FMLP in concentrations too low to stimulate the formation of O_2^- itself enhanced the production of O_2^- elicited subsequently by stimulation with concanavalin A, opsonized zymosan, or phorbol myristate acetate [27].

As a non-phagocytosable, non-protein stimulus detergents have been used, especially digitonin [28,29] but also sodium dodecyl sulfate [30]. Fatty acids which elicit the burst include myristate [31] and arachidonate [32,33]. Active lectins include concanavalin A and phytohemagglutinin [34]; the specific synthetic inhibitor, alpha methyl mannoside, blocked both binding of concanavalin A and the concanavalin A-induced stimulation of respiration, suggesting that concanavalin A acted through its known specificity for oligosaccharides.

Continuous stimulation with concanavalin A was considered necessary because the burst could be blunted by the addition of alpha methyl mannoside after Con A stimulation had already commenced, suggesting, that, displacement of Con A from its receptor could reverse the stimulation. The fungal metabolites, the cytochalasins, exert varied effects; whereas cytochalasin B enhances the formation of O_2^- by other agents, especially particulate stimuli which can be internalized [18,23-25], cytochalasin D [35,36] and E [37] appear to stimulate the formation of O_2^- directly. The ability of A 23187 to elicit formation of O_2^- from PMNs [24,59,60,68,70] appears to be due to its action as an ionophore for Ca (see below). Other active stimuli include fluoride, notable because its effect is reversible [38], human leucocyte pyrogen which might have a function in vivo [39], and phorbol myristate acetate [40,41]. The latter has been very useful as a potent stimulus of the burst in various studies of the kinetics and mechanism of activation of the burst.

2.2 Agents which can Block the Elicitation of the Burst

In addition to the specific antagonist, alpha methyl mannoside (of Con A) and carbobenzoxy phenyl alanyl methionine (of FMLP), other compounds which inhibit the burst include inhibitors of proteases, adrenal steroids, anti-inflammatory drugs, phenothiazines, and local anaesthetics. Simchowitz et al. [42] inhibited production of O_2^- by PMNs stimulated with FMLP by the action of the inhibitors of proteases, tosyl phenyl alanyl chloromethyl ketone, tosyl lysyl chloromethyl ketone, and the methyl esters of phenyl alanine and tryptophan. The effects of these, but not other compounds were thought to indicate that the putative protease was more similar to chymotrypsin than to trypsin. These inhibitors of proteolysis had little effect on the Ca^{2+} influx which accompanies activation, prompting the interpretation that the inhibitors act after the influx of Ca^{2+} (see section on "Signal Transmission"). The inhibitory effect of di-isopropylfluorophosphate (DFP), surprisingly, was reversible, an observation not consistent with the covalent modification of serine residues in a serine protease, its usual mode of action. Kitagawa et al. [43] similarly found that a variety of inhibitors of proteases inhibited the formation of O_2^- elicited by Con A or cytochalasin E. These authors found inhibition by compounds with both tryptic and chymotryptic specificity. Inhibition was much greater when the inhibitor was added before the stimulus. Goldstein et al. found an even broader range of inhibitors of proteases to reduce the formation of O_2^- by PMNs including several natural inhibitors: soybean trypsin inhibitor, lima bean trypsin inhibitor and antipain [44]. These authors encountered a pitfall, perhaps not detected in some other studies, namely the inhibition by DFP of the reduction of cytochrome c by an *enzymic* source of O_2^-. If DFP can interfere with the reduction of cytochrome c by O_2^- rather than inactivate the enzymic source of O_2^-, the nature of the reversible effect of DFP in the experiments of Simchowitz et al. might be explained.

The local anaesthetics, tetracaine and lidocaine, inhibited both the release of lysosomal enzymes and the formation of O_2^- by PMNs [45] but with variable cytotoxicity as judged by release of lactic dehydrogenase, a cytoplasmic enzyme. Tetracaine, which appeared to be less toxic, reduced the formation of O_2^- by PMNs stimulated with either concanavalin A or C_{5a} (generated by the action of zymosan on serum in the presence of epsilon amino caproic acid). The slight effects of tetracaine on binding of ^{3}H-Con A at 37° were felt to be insufficient to account for the extent of inhibition observed.

Adrenal cortical steroids reduce the response of PMNs to stimulation by serum-treated zymosan, heat aggregated IgG, C_{5a} [46] or FMLP [47]. Steroids diminished both release of lysosomal enzymes and formation of O_2^-. These effects of adrenal steroids were suggested as possible explanations for their antiinflammatory effects in vivo. This is supported by the observation that similar effects are seen in PMNs from human subjects to whom steroids were administered [48]. One explanation for the inhibitory effect of glucocorticoids on the metabolism of PMNs was suggested by the work of Hirata et al. [49] who found that the diminished chemotaxis of rabbit PMNs to FMLP was associated with reduced release of incorporated ^{14}C arachidonic acid from PMNs in response to FMLP. This diminished release was attributed to inhibition of phospholipase A_2 by a protein of molecular weight of 40,000 daltons which was synthesized in response to administration of glucocorticoids. How this might explain the inhibition of the formation of O_2^- is not clear but it might be related to the ability of exogenous arachidonate acid to stimulate the formation of O_2^- by PMNs [32] and of arachidonic acid to stimulate the influx of Ca^+ in PMNs, [33] a step which can activate the formation of O_2^- (see below).

The actions of non-steroidal anti-inflammatory drugs appear to be diverse. Phenylbutazone and sulfinpyrzone inhibited several effects of FMLP on PMNs: increased adhesiveness, stimulation of the hexosemonophosphate shunt, lysosomal enzyme release, and formation of O_2^- [50]. These effects were explained by the ability of both phenylbutazone and sulfinpyrazone to inhibit binding of labelled FMLP to PMNs. Both were selective in that neither inhibited responses of PMNs to C_{5a} and neither inhibited the stimulation of the hexose monophosphate shunt by latex or by opsonized candida. The responses of PMNs to FMLP, including the release of O_2^- have also been inhibited by 5, 8, 11, 14 eicosatetraynoic acid, an inhibitor of the metabolism of arachidonic acid [51,52]; by indomethacin [47,52,53]; and by bromophencyl bromide, an inhibitor of phospholipase A_2 [52]. Using serveral stimuli including FMLP, zymosan-activated serum, serum-treated zymosan, A 23187, and a complex of bovine serum albumin with anti-bovine serum albumin, Smolen and Weissman [52] found the extent of inhibition by these three agents to vary with the stimulus. Other inhibitors of the formation of O_2^- by PMNs include ibuprophen [47,53], oxyphenbutazone, mefenamic acid and benzydamine [53].

The ability of one stimulus to induce refractoriness to a second challenge with the same stimulus is called deactivation. This process appears to be specific for each stimulus [54]. Exposure of PMNs to FMLP diminished the formation of O_2^- and blocked the increase in the concentration of cAMP on second challenge with FMLP. However, both responses were elicited normally by C_{5a}. The reverse was also true. Whether this phenomenon is due to internalization or occupancy of receptors for one stimulus while leaving intact those for the other is not known. The basis for the inhibitory effects of endotoxin on the responses of PMNs [55] is unclear. For the inhibitory effects of phenothiazines see "Transmission of Signal."

2.3 Receptors

The receptors for the various stimuli appear to be the same for some chemically related stimuli. Becker has presented compelling arguments that the various synthetic peptides act at the same receptor [56]. His argument is based on:

1) the same hierarchy of potency for the various peptides whether one measures chemotaxis, release of lysosomal enzymes, formation of O_2^- or aggregation; and
2) the same K_D for the competitive inhibitor, carbobenzoxyphenyl alanyl methionine, with all four responses of PMNs as well as by inhibition of binding of the peptides.

Bacterial neuraminidase inhibits the formation of O_2^- by PMNs challenged with latex particles or with Con A, suggesting an essential sialic acid in these receptors [57]. Treatment with neuraminidase also caused PMNs to form less H_2O_2 in response to *Staph. aureus* and to kill fewer of the bacteria [58]. The blockade of the effects on Con A by alpha methyl mannoside has already been mentioned.

2.4 Transmission of the Signal from the Surface of the Cell

Several steps in the transmission of the signal from the surface of the cell to the site of formation of O_2^- appear to occur. Several groups have presented evidence that a change in membrane potential is an early event following stimulation of PMNs [59-61]. Whitin et al. [61] noted a change in fluorescence of 3,3′-dipropylthiocarbocyanine within 10 s of the onset of formation of O_2^-. The increase in fluorescence of the carbocyanine dye, a lipophilic cation, was measured as a reflection of the increase in membrane potential. The close correspondence of changes in membrane potential and the subsequent initiation of the formation of O_2^- was further demonstrated by showing that two agents which affected the activation of the formation of O_2^- in known ways also affected the depolarization in a similar fashion. Incubation with 2 deoxyglucose (1 mM) completely abolished the change in fluorescence. Chlorpromazine (20 μM) had similar inhibitory effects on the lag time both for the formation of O_2^- and for the onset of depolarization of the membrane. It also inhibited the formation of O_2^- and the extent of the change in membrane potential to a similar degree, 88% and 78% respectively. Finally, in PMNs from individuals afflicted with the inherited defect in the formation of O_2^- called chronic granulomatous disease, no depolarization occurred in response to PMA.

Using both a carbocyanine dye and another lipophilic cation, triphenylmethylphosphonium, Seligmann and Gallin [60] found defective changes in the membrane potential in PMNs from patients with chronic granulomatous disease in response to FMLP or PMA. In PMNs from patients with other disorders of function the responses were normal. The resting membrane potential, the response to valinomycin, (an ionophore for potassium) and the response to the ionophore for Ca^{2+}, were all normal or only slightly impaired.

Korchak and Weissmann [59] employed triphenylethylphosphonium as a means of exploring the effects of external sodium on the responses of PMNs to stimulation with an antigen-antibody complex (bovine serum albumin-antibovine serum albumin), an ionophore for calcium (A 23187), and Con A. Both the formation of O_2^- by PMNs and the release of their lysosomal enzymes were diminished by reduction in the concentration of sodium in the medium. The uptake of the lipophilic cation was also reduced by the reduction in concentration of sodium indicating an effect on membrane potential. To complicate matters, the effects of an ionophore for Ca^{2+} were also diminished by reduced concentration of Na^+ in the medium, and an ionophore for Na^+ failed to stimulate the formation of O_2^-. The authors postulate a role for Na^+ in the binding of ligand to receptor and also subsequently in the events leading to the formation of O_2^-. In support of these data suggesting an interaction of the effects of Na^+ and Ca^{2+} are the data of Simchowitz and Spilberg [62] who demonstrated a unidirectional interaction between the fluxes of Ca^{2+} and Na^+ in PMNs in response to FMLP. Whereas the uptake of ^{22}Na was not affected by the concentration of Ca^{2+} the influx of Ca^{2+} decreased as the concentration of Na^+ was reduced. On the basis of these data they suggest that the influx of Na^+ precedes that of Ca^{2+}.

The participation of Ca^{2+} in the activation of the PMN to form O_2^- has been demonstrated in a number of ways. In the absence of Ca^{2+} in the medium, the formation of O_2^- in response to FMLP is markedly reduced [24,25,63]. The ionophore for Ca^{2+} A 23187 stimulated the formation of O_2^- [24,59,60,68,70]. The chelator of Ca^{2+}, EGTA, reduced the formation of O_2^- [29].

However, the effects of Ca^{2+} are not seen with all stimuli; the production of O_2^- in response to phorbol myristate acetate was unaffected in medium lacking Ca^{2+}. Lanthanum, which can act as a competitive inhibitor of Ca^{2+}, reduced the effect of Ca^{2+} in suspensions of PMNs stimulated with FMLP [63]. That Ca^{2+} acted by entering the cell was suggested because calcium ions did not affect the binding of FMLP. However, verapamil which blocks the channel for Ca^{2+} in cell membranes, produced dose-dependent inhibition of the formation of O_2^- and of the influx of ^{45}Ca. This inhibition could be overcome by increasing the concentration of Ca^{2+} in the medium [63]. The reasons for the requirement of Ca^{2+} for activation by some stimuli but not by others might be related to the site at which the stimulus in question acts. PMA, for example, might act at a site beyond the point at which Ca^{2+} is required.

The reduction in the fluorescence of chlortetracycline which occurs when it binds Ca^{2+} has also been used to show that changes in the intracellular concentration of Ca^{2+} occur when the formation of O_2^- is stimulated in PMNs. When PMNs which had been treated with chortetracycline were stimulated with either cytochalasin D or *E. Coli*, a decrease in fluorescence occurred which followed closely the release of O_2^- [64]. The antagonist of Ca^{2+}, TMB8 inhibited both the change in fluorescence and the formation of O_2^- by PMNs, leading the authors to conclude that translocation of Ca^{2+} was required for the activation of PMNs by these two stimuli.

Finally, the inhibitory effects of phenothiazines [65] on the formation of O_2^- by PMNs have been interpreted as evidence for the participation of calmodulin in the transmission of the signal from the surface of the cell to the oxidase. One action of phenothiazines is to inhibit the effects of the complex of calmodulin and Ca^{2+}. The activity of the effector molecule, such as phosphodiesterase, which would normally be stimulated by the binary complex of Ca with calmodulin, is not stimulated in the presence of phenothiazines. Jones et al. [65] have shown that in both intact PMNs and in membranous fragments of PMNs stimulated with opsonized zymosan, phenothiazines inhibit the formation of O_2^-. The hierarchy of potency of the various phenothiazines is the same as the hierarchy of potency in preventing the stimulation by calmodulin of the action of phosphodiesterase. Moreover, the concentrations of the phenothiazines which were effective were very similar to the concentrations required to block the effect of calmodulin on phosphodiesterase. When membranous fragments made from PMNs stimulated with opsonized zymosan were titrated with EGTA in a medium containing calcium, maximal inhibition of the formation of O_2^- from NADPH occurred below 2 μM Ca but rose very rapidly at slightly greater concentrations. This observation indicated that the activity was affected over a very narrow range of concentrations of calcium, a salient characteristic of reactions modulated by the complex of calmodulin with Ca^{2+}. In membranous preparations from PMNs stimulated with opsonized zymosan calmodulin stimulated the formation of O_2^- from NADPH. Finally, when a particulate subcellular fraction from PMNs activated with FMLP was prepared in the presence of EDTA and resuspended in buffer containing Ca^{2+}, no activity of the oxidase was seen, but marked stimulation with calmodulin occurred, restoring the activity to that seen when the same cells were prepared in the absence of EDTA. These data suggested that preparation of subcellular fractions in the presence of EDTA removed some activity necessary for full activation of the oxidase and that calmodulin in the presence of Ca^{2+} fully restored the activity. However, efforts to elicit the activity of the enzyme forming O_2^- with calmodulin and Ca^{2+} in 'resting' PMNs have not been successful [65)].

However, another interpretation of the effects of phenothiazines is based on the same hierarchy of potency of different members of the phenothiazine family in inhibiting hemolysis and anaesthetizing nerves as in suppressing the formation of O_2^- by PMNs. In these situations phenothiazines were proposed to render the membrane more fluid and to increase the area of the membrane. By analogy, phenothiazines might act in PMNs by increasing the fluidity of the membrane, thereby preventing the formation of a stable complex of separate components required for activity of the oxidase. Conversely, the authors propose that in the normal process of activation membrane depolarization occurs, followed by a change (presumably a decrease) in fluidity of the membrane which enhances the likelihood of a stable association of the components of the active complex [66)]. It might seem that an increase rather than a decrease in fluidity might promote mobility in the lipid bilayer and thereby increase the likelihood of association of the individual components. Decreased fluidity, however, could stabilize a macromolecular complex.

Another action of phenothiazines is to compete with NADPH for the oxidase, an inference which was based on fulfillment of the criteria for competitive inhibition in double reciprocal plots of 1/[NADPH] vs. 1/rate of formation of O_2^- in the presence and absence of inhibitor [66)]. It seems worth considering that all the effects of phenothiazines might be mediated through this effect and that the process of activation represents the presentation of substrate to the enzyme from which, in the resting state, the substrate is kept separate.

At least two groups have tested the effects of blockade of the anion channel on the production of O_2^- by PMNs [67, 68)]. Gennaro and Romeo [67)] found that the production of O_2^- by bovine PMNs was reduced by the blockers of the anion channels SITS (4-acetamido-4'-isothiocyanostilbene-2,2'-disulfonic acid) and ANS (anilino naphthalene disulfonic acid). The consumption of O_2 was unaffected. In accord with the expectation if the effects of these agents were to block the anion channel, replacement of the chloride in the medium with salicylate, thiocyanate, and nitrate inhibited the formation of O_2^- in the expected hierarchy (salicylate > SCN^- > NO_3^- > Cl^-). The effects of these alternate anions, however, was small. These authors interpreted their data to mean that superoxide was formed within the PMN and exited into the medium through the anion channel [67].

A different conclusion was reached by Korchak et al. [68] in studies of human PMNs. Blockers of the anion channel reduced the secretion of lysosomal enzymes in response to immune complexes and the ionophore for Ca^{2+}, A 23187, but did not affect generation of O_2^-. Whether the differences in the results of the two studies are wholly attributable to the different species studied (bovine vs. human) or to other differences is not clear.

Evidence has been presented that the concentration of cAMP is transiently increased in PMNs stimulated to form O_2^- by FMLP or C_{5a} [69,70]. Paradoxically, agents which increase the intracellular concentration of cAMP caused a dose dependent fall in the formation of O_2^- by PMNs which was elicited by FMLP. A rise in the concentration of cAMP within PMNs was also observed by Smolen [70] after stimulation with several agents. The increase in the concentration of cAMP preceded the release of lysosomal enzymes and the formation of O_2^- but did not occur before the change in membrane potential. The increase in the concentration of cAMP was judged not to be sufficient for the elaboration of O_2^- or the secretion of enzymes from lysosomes because three maneuvers produced changes in cAMP but no subsequent response. However, whether the concentration of cAMP must rise in the normal chain of events leading from stimulus to formation of O_2^- is not clear.

A role for microtubules and microfilaments was proposed by Nakagawara and Minakami [71] based on their observations that formation of O_2^- elicited by cytochalasin E was stimulated by D_2O and inhibited by colchicine, vinblastine and cAMP.

2.5 Activation of the Oxidase

There is very little information on the process by which the enzyme(s) which catalyze the formation of O_2^- is activated. Whether activation represents covalent modification of a proenzyme, translocation to the subcellular site where it gains access to its substrate(s), insertion of a necessary cofactor, association of active subunits, or still another process remains to be determined. The rapidity with which activation occurs seems too rapid to represent synthesis of protein de novo.

2.6 The Oxidase

2.6.1 Location

The inference that the oxidase(s) is an ectoenzyme has been made from a variety of data. The demonstration that PMNs possess both the Cu/Zn and the Mn superoxide dismutase [72-75] prompted the suggestion [72] that O_2^- must be formed at a site which protected it from the action of these enzymes. Since O_2^- appeared in the fluid surrounding the PMN, the most advantageous site appeared to be the plasma membrane. In accord with this proposal, treatment of PMNs with a nonpenetrating organic anion, p-diazobenzene sulfonic acid, in association with cytochalasin B to prevent internalization of membrane into a phagocytic vacuole, inhibited the generation of O_2^- [76]. The binding of the stimulus, Con A was not inhibited, prompting the interpretation that the non-penetrating reagent had not interfered with binding of the stimulus but instead had inactivated the oxidase exposed at the surface of the PMN. Similar conclusions were reached with the non-penetrating reagent p chloromercuribenzoate [77].

Fractionation of subcellular particules of PMNs has yielded data which were interpreted diversely. Baehner [78] fractionated lysates of PMNs from infected humans by differential centrifugation and by centrifugation on sucrose-density-gradients and found the maximal specific activity of an enzyme which reduced NBT with NADH (or NAPDH) as substrate to be enriched in the fractions with the maximum specific activity of acid and alkaline phosphatases. The differing location of the NADH-NBT oxidoreductase in the sucrose gradient from the location of the enzymes in granules and the lack of latency (rise in activity with the addition of Triton X-100) of the oxidase in contrast to the latency of enzymes in the granules (peroxidase, beta-blucuronidase, and lysozyme) led to the conclusion that the NADH-NBT oxidoreductase was not associated with the granules and was located either in the plasma membrane or in the microsomes. Whether these studies actually identified the oxidase responsible for the formation of O_2^- is problematical since activation in vitro is necessary for the full expression of this activity. Whether the increased reduction of NBT which probably characterized the PMNs from these infected patients meant that the oxidase was fully activated in vivo seems doubtful. Whether the reduction of NBT was due to the formation of O_2^- in these studies is also not clear.

In another approach subcellular fractionation of human PMNs, stimulated in vitro with phorbol myristate acetate or opsonized zymsan, was performed by rate-zonal centrifugation in sucrose gradients [79]. By varying the centrifugal force the NADPH oxidase could be separated from the activities of enzymes in the lysosomal granules but was similar to the distribution of markers for the plasma membrane, aryl phosphatase, alkaline phosphodiesterase I, and acid aryl phosphatase.

Cohen et al. [80] prepared podosomes (membrane-containing fractions from lightly sonified PMNs) from the digitonin-stimulated PMNs of guinea pigs. The effects of pH, N-ethyl maleimide, glucose, CN^-, 2-deoxyglucose, and Ca^{2+} on the generation of O_2^- by intact cells and on the formation of O_2^- in the podosomes were all the same, leading to the interpretation that the activity found in the podosomes was the activity in whole cells responsible for the generation of O_2^-. The podosomes were also enriched in adenylate cyclase and 5'-nucleotidase, markers of the plasma membrane. The K_m for NADPH was $^1/_{10}$ the K_m for NADH. The increased specific activity of the oxidase from PMNs activated in the presence of 5 mM glucose and 0.5 mM CN^- was attributed to activation of greater amounts of the same enzyme found in the absence of glucose and CN^- since the K_ms for NADPH and NADH were unchanged while the specific activity increased two-fold. The data support localization of the activated oxidase in the plasma membrane.

In a second study by the same group [81] phagocytic vacuoles were prepared from PMNs which had been exposed to opsonized oil particles. The phagocytic vesicles could be collected because of the distinctive buoyant density in a discontinuous gradient of sucrose of the oil within phagocytic vacuole. The effects of CN^-, of chelators, and of EDTA on the appearance of activity in the phagocytic vacuoles were similar to those effects on the intact cells. The activity of the oxidase in the phagocytic vacuoles in isolation fell while other activities in these particles (5′ nucleotidase and lysozyme) did not. This indicates that the loss of activity of the oxidase in whole cells was not simply internalization of the membrane, thereby sequestering O_2^- where it can no longer reduce cytochrome c, nor destruction of all the enzymes within the phagosome indiscriminantly.

An oxidase has also been localized to the surface of human PMNs by cytochemical staining [82]. Cells were allowed to phagocytose polystyrene spheres in the presence of Ce^{3+} and the site of formation of H_2O_2 was detected by locating electron-dense deposits of $Ce(OH)_2OOH$ by electron microscopy. The reaction product was found both in phagocytic vacuoles and at the cell surface in the presence of NADH. Its formation was diminished by catalase and was dependent on NADH in the medium. Tetrazolium salts inhibited the formation of the reaction product at the cell surface, suggesting that when they served as the recipient of electrons, they competed with dioxygen successfully and prevented the formation of H_2O_2. The deposition of formazan, the insoluble product of the reduction of tetrazolium salts, was not inhibited by superoxide dismutase, suggesting to the authors that the product of this reaction was not O_2^-. The interpretation of these data is difficult since the formation of the product was slight in the absence of exogenous NADH and yet the burst occurs in the absence of added pyridine nucleotide. The lack of inhibition by superoxide dismutase might be a consequence of the inability of the negatively-charged enzyme from bovine erythrocytes (pI 4.8) to approach the negatively-charged membrane of the cell to compete successfully with the tetrazolium salts for O_2^-.

Some of the other studies in which the oxidase was proposed to be located in lysosomal granules [83] may be difficult to interpret because the conditions of the assay for the oxidase at pH 5.5 or 6.0 allowed a large component of a chain reaction propagated by free radicals [84]. The association of activity in the formation of H_2O_2 with the granules of PMNs was based on the presence of myeloperoxidase in the subcellular fraction which was sedimented at 13,000 g [85]. Although this centrifugal force might not have been expected to sediment the plasma membrane, the stimulation of the PMNs by heat-killed *E. Coli* could well have caused the phagosomes to sediment more rapidly than isolated membranes. Since other markers were not determined the purity of the fraction identified as granular is unclear.

2.6.2 The Reaction which the "Oxidase" Catalyzes

In a number of earlier studies the oxidation of NADH or NADPH was assayed under conditions (acid pH, Mn^{2+}) in which there was a large component of what appears to be a chain reaction propagated by free radicals [86-94]. Under these conditions it is difficult to be certain how much of the formation of O_2^- and of the oxidation of pyridine nucleotides was due to turnover of the enzyme and how much was due to the chain reaction.

A variety of data have suggested that either NADH or NADPH are the substrate for the oxidase. Thoughtful commentaries on this apparent conflict have appeared [95,96]. Some of the apparent discrepancies may well arise from different groups of investigators studying PMNs from different species under widely divergent conditions.

The substrate for the oxidase which forms O_2^- in PMNs appears to be NADPH, at least in human PMNs. The data in support of this contention include:

a) a fall in the ratio of NADPH to NADP after stimulation with polystyrene spheres while the ratio of NADH to NAD remains the same [97];
b) the existence of a defect in the formation of H_2O_2, bacterial killing, and activation of the HMP in individuals with severe deficiency of glucose 6 phosphate dehydrogenase, a condition which results in an insufficient supply of NADPH but not NADH [98];
c) a lesser K_m for NADPH than for NADH in the partially purified preparations of the oxidase from activated human PMNs [99, 100, 81]; and
d) stimulation of the hexose monophosphate shunt which yields NADPH but not NADH, during phagocytosis [101].

Although the assay of the oxidase in PMNs from patients with chronic granulomatous disease of childhood in whom formation of O_2^- is markedly depressed, would seem a further important criterion, the activity has either been found to be markedly decreased with both substrates [100] or in resting PMNs has been found to be decreased [102] or normal [103] with NADH as substrate. Since the defect in CGD is expressed after stimulation rather than in the resting state, the relevance of the data in resting PMNs is unclear. The demonstration that formation of H_2O_2 by PMNs from normal individuals but not from patients with chronic granulomatous disease was stimulated by external NADH but not NADPH [104] is problematical because of the difficulty of interpreting the relevance of the effects of external pyridine nucleotides.

The reaction in which O_2^- is formed. The reaction in which O_2^- is formed has been suggested to be:

$$NADPH + 2\,O_2 \rightarrow NADP^+ + H^+ + 2\,O_2^- . \tag{7}$$

If the O_2^- subsequently dismutes,

$$O_2^- + O_2^- + 2\,H^+ \rightarrow H_2O_2 + O_2 , \tag{8}$$

the sum of these reactions is:

$$H^+ + NADPH + O_2 \rightarrow NADP^+ + H_2O_2 . \tag{9}$$

Attempts to measure the ratio of the oxygen consumed to O_2^- formed to H_2O_2 produced in intact cells have been difficult because of the many reactions in which both O_2^- and H_2O may participate. When PMNs were treated with azide to inhibit the reactions of H_2O_2 with the hemoproteins, catalase and myeloperoxidase, and with cytochalasin B to prevent formation of phagocytic vacuoles in which O_2^- might be inaccessible to the cytochrome c used to detect it [105], the ratio of O_2 consumed to O_2^- formed to H_2O_2 produced was 1.0:0.99:0.4. The interpretation of these ratios is difficult indeed. Hydrogen peroxide may be acted on not only by catalase and myeloperoxidase but also by glutathione peroxidase. How many other reactions O_2^- and H_2O_2 may enter is not clear, rendering the ratio to be expected uncertain. The complexities of the interpretation of these data have been pointed out [96].

2.6.3 The Oxidase Itself

The enzyme chosen for isolation clearly will have different properties depending on which of its characteristics are inferred from studies of intact cells. Most studies have concentrated on a particulate enzyme detectable only in cells stimulated to form O_2^-, which uses NADPH as substrate [99, 106–110]. Enzymes with other characteristics have also been partially purified and studied as well [117, 118].

Babior, who has studied this enzyme at several stages of its purification, found in lysates of PMNs which were activated with zymosan that of eight potential biological reductants only reduced pyridine nucleotides supported the formation of O_2^-. The K_m for NADPH was less than the K_m for NADH and the activity was decreased in preparations from three patients with chronic granulomatous disease. In accord with predictions based on reaction 7, 0.55 molecule of O_2^- was measured per molecule of NADPH oxidized under conditions of saturating concentrations of cytochrome c [100]. The enzyme which was extracted with Triton X-100 from a granule-rich fraction from activated PMNs, required an external source of FAD for the formation of O_2^- from NADPH [106]. Riboflavin and FMN would not substitute. Flavin adenine dinucleotide was proposed as a necessary cofactor, which was probably lost when the enzyme was treated with the detergent.

The enzyme, which was extractable from activated but not from resting PMNs, exhibited a quadratic dependence on enzyme concentration [107] which could be restored to a linear relationship by the addition of phosphatidyl ethanolamine. This phospholipid may have been removed by treatment with detergent. The K_m of the enzyme in Triton X-100 in the presence of FAD was greater for NADH (930 μM) than for NADPH (33 μM) and the optimal pH was broad with a maximum at 7.0 [99]. Its activity in the presence of phosphatidyl ethanolamine was judged sufficient to account for all of the O_2^- which activated PMNs produced.

The enzyme was found to be unstable [110] in soluble form and to bind FAD tightly with a K_m of 61 μM. Its selectivity for an acceptor of electrons was demonstrated by its failure to reduce ferricyanide, dichlorophenolindophenol, methylene blue, or cytochrome c (directly). Tantalizingly the loss of activity was nearly immediate in the presence of ATP, prompting the suggestion that ATP may be a physiological regulator of the activity of the enzyme. Activity was stimulated by dithiothreitol at low concentrations and loss of activity was stimulated by salts and EDTA.

In a similar approach the "oxidase" was extracted with deoxycholate in 12–14% yield from the granules prepared from PMNs which had been activated with either phorbol myristate acetate or with opsonized zymosan [108]. By this method, as well, the enzyme was very unstable and its specific activity in deoxycholate was only 0.1–0.2 of the activity in the particulate fraction from which it was extracted. The spontaneous loss of activity was slowed at 4° and in the presence of glycerol, dimethyl sulfoxide, and ethylene glycol. Further studies by the same group [109] provided the surprising observation that the ratio of moles of NADPH oxidized to O_2^- produced varied with the stimulus used to activate the PMNs, being greater in lysates of cells stimulated with PMA. In contrast, the effects of phorbol myristate acetate and of opsonized zymosan on consumption of O_2, on formation of O_2^-, and on activation of the hexose monophosphate shunt were similar in the intact cells. The enzyme was found to undergo autoinactivation, loss of activity which depended on enzymic turnover. Deoxycholate both stimulated activity and accelerated autoinactivation. Autoinactivation could not be attributed to exhaustion of substrate nor to attack by the reactive derivatives of molecular oxygen which would be expected to result from O_2^-, since scavengers of O_2^-, H_2O_2, $OH^\cdot$, and $^1O_2$ all failed to retard the loss of activity. The rate of oxidation of substrate fell markedly when the deuterated substrate, 4R-(^{2}H)-NADPH, but not when 4S-(^{2}H)-NADPH was substituted for the hydrogenated substrate, indicating stereospecificity of hydride transfer. As with the results of Babior and Kipnes [106] FAD stimulated the activity of the enzyme in Triton X-100 and titration with FAD suggested that in Triton X-100 the enzyme was 40–75% depleted of flavin. Substitution of synthetic flavins showed that an analog only capable of transferring two electrons at a time inhibited the enzyme at a concentration of 32 μM, suggesting that the acceptor of electrons will only accept one electron at a time. Finally, gel filtration of the enzyme in detergent resulted in enrichment in a chromophore with spectral characteristics of a cytochrome b but depletion of myeloperoxidase, possibly supporting the earlier suggestion (see below) that cytochrome b is incorporated into the phagosome. The b-type cytochrome was only slowly reduced in the presence of NADPH but it did react briskly with O_2.

The role of cytochrome b in the formation of O_2^- by PMNs was initially suggested by experiments in which a chromophore was detected in subcellular fractions of PMNs, the spectral characteristics of which were felt to be distinct from those of myeloperoxidase, another hemoprotein abundant in PMNs [111]. The subcellular distribution of this pigment by contrifugation in gradients of sucrose suggested that it was located in two different fractions, one of which was thought to be the granules and the other the plasma membrane as demonstrated by cosedimentation with ^{3}H-Concanavalin A [112]. The ability of dithionite, which penetrates membranes slowly, to reduce the cytochrome b was thought further to support its location in the plasma membrane as was the rapid appearance of the pigment in phagocytic vaculoes [113]. The function of this cytochrome was thought to be intimately involved in the respiratory burst since PMNs from patients with chronic granulomatous disease in whom the respiratory burst is markedly reduced were found to have decreased amounts of the pigment [114,115]. This has not been confirmed by another group who found the problem confounded by the presence of eosinophil myeloperoxidase whose absorbance maxima would obscure the identification of cytochrome b [116]. In the latter report females affected with chronic granulomatous disease were found not to have reduced amounts of cytochrome b but the examination of males, the usual patients with chronic granulomatous disease, was not possible due to the contamination with the myeloperoxidase from eosinophiles. While the assay of the males with chronic granulomatous disease was rendered inconclusive, other data suggested that the cytochrome b was not involved in the formation of O_2^-. By treatment with Brij 35, a detergent, the cytochrome be could be studied in isolation; it did not bind CO, indicating that it is not an O-type cytochrome. Finally, removal of cytochrome b from the NADPH-dependent oxidase, had no effect on the ability of the latter to form O_2^-, strongly suggesting at least with this enzyme, that the cytochrome b is not involved in the formation of O_2^-.

An oxidase has been purified from the resting PMNs of guinea pigs which appears to be far more active with NADH as substrate than with NADPH [117]. This enzyme, which was prepared in isotonic KCl rather than in detergent, donated only about 15% of the electrons removed from its substrate to O_2 to form O_2^-. The rest were donated directly to O_2 in pairs to form H_2O_2. The ratio of H_2O_2 formed, whether directly or indirectly by dismutation of O_2^-, to NADH oxidized was 0.95–0.99. The enzyme was inhibited by a variety of phosphates, including ATP and ADP, the extent of inhibition increasing with the number of phosphates. The inhibitory effects of ATP were shown to be exerted through competition with NADH and could be reversed by the addition of Mg^{2+}. The catalytic turnover coupled with the concentration of NADH in PMNs of guinea pigs were sufficient to account for the respiratory burst. The high K_m (0.4 mM) was recognized and various explanations were suggested as possible means for modulating the affinity for substrate. These included a change in substrate affinity when the enzyme associates with the inner face of the plasma membrane, the site at which the authors proposed that the enzyme resides based on the substantial proportion which remains associated with the plasma membrane when cells are disrupted in isotonic sucrose rather than in KCl.

Another oxidase was isolated from the resting PMNs of guinea pigs by the same group which isolated the NADH oxidase [118]. In contrast to the NADH oxidase this enzyme, which uses a variety of aldehydes as well as 2 hydroxy pyrimidine as substrates, donates a larger proportion of its electrons to O_2 to form O_2^- (75–90%). The apparent molecular weight was 340,000 by gel filtration and the optimal pH was 7.5 to 9.0. In spite of its localization in the phagosomal membrane by electron microscopic histochemistry, the enzyme was thought not to represent the dominant source of O_2^- in PMNs from guinea pigs because it was inhibited by CN^- whereas the respiratory burst is not. It was not found in human PMNs. There are attractive properties to this enzymic reaction since one action of myeloperoxidase on amino acids is to form aldehydes which would be substrates for this enzyme.

2.7 Termination of the Burst

At least two types of mechanism for termination of the respiratory burst have been proposed. The participation of myeloperoxidase has been suggested from observations in PMNs from patients with the hereditary deficiency of myeloperoxidase and from studies with inhibitors. Rosen and Klebanoff [119] reported that the rate of formation of O_2^- declined less rapidly in suspensions of PMNs from patients with the inherited deficiency of myeloperoxidase. Jandl et al. [120] described the effects of two types of inhibitors on the termination of the respiratory burst. The effect of azide was on the duration of production, but not on the maximal rate of formation, of O_2^-. Azide was thought to act by inhibiting myeloperoxidase which could inactivate the oxidase either by forming products such as HOCl, which attack the oxidase, or by direct action of myeloperoxidase on the oxidase acting as the reducing substrate. The decline in activity of the burst was also inhibited by cytochalasin B, prompting the interpetation that part of the apparent loss of activity of the oxidase was caused by internalization of the enzyme, possibly in the phagocytic vacuole. A somewhat puzzling observation was that when PMNs were activated under N_2 the subsequent formation of O_2^- by particles prepared from the PMNs was markedly enhanced compared with particles from PMNs activated in the presence of air. When the stimulation of the PMNs was continued for 60 min under N_2, however, the production of

O_2^- had fallen considerably in particles prepared from the anaerobically stimulated cells. Whether a third type of inactivation was the explanation for these observations is not clear. Certainly the very low K_m of the oxidase of O_2 of 10 μM [121] might indicate that only the most rigid measures would exclude O_2 sufficiently to abolish the enzymic turnover in PMNs. Therefore, the partial effect of anaerobiosis might have been to slow both enzymic turnover and the consequent autoinactivation found by Light et al. [109] in the partially purified enzyme.

2.8 Products and Consequences of the Burst

Although PMNs appear to reduce oxygen in the respiratory burst by a single electron, to form O_2^-, both $OH^\cdot$ and H_2O_2 may be produced from subsequent reactions as summarized earlier. The effects of the formation of these reactive molecules range from the generation of light to the killing of other cells.

The evidence that O_2^- is a product of the burst has already been summarized. Another direct consequence of the formation of O_2^- is the reduction of nitroblue tetrazolium to form an insoluble deposit of blue formazan. The reduction of nitro blue tetrazolium had been observed to occur in PMNs in patients with bacterial infections. In intact cells this appears to be a consequence of the formation of O_2^-; it fails to occur in PMNs from patients with chronic granulomatous disease which cannot undergo the respiratory burst [123], it is diminished by anaerobiosis [122], and it is inhibitable by superoxide dismutase [123, 124].

The many methods by which the formation of H_2O_2 by PMNs has been measured have been summarized in the superb volume by Clark and Klebanoff [9]. At least one other method has been successfully employed in which the spectrophotometric reagent which forms a characteristic complex with H_2O_2 is a peroxidase. In this method, a peroxidase is added to the medium in which PMNs or their subcellular fractions are stimulated to undergo the respiratory burst. In the absence of the reducing substrate for the peroxidase the enzyme is "arrested" in its catalytic turnover after it has been oxidized by H_2O_2. The heme iron in this oxidation state exhibits a characteristic absorption spectrum. From the change in absorbance the amount of H_2O_2 formed can be determined [85, 125].

Formation of the hydroxyl radical. Because the killing of certain microbes was dependent on the ability of PMNs to undergo the respiratory burst, the actual microbicidal species was sought. It had been shown that a potent oxidant was formed during the catalytic action of xanthine oxidase on xanthine [126], an enzymic reaction which, like the PMN, produces both O_2^- and H_2O_2. This potent oxidant was proposed to be the hydroxyl radical formed by the reaction between O_2^- and H_2O_2 (reaction 4). Because such an oxidant seemed a likely candidate to mediate the microbicidal activity of PMNs, the formation of $OH^\cdot$ was assessed in PMNs.

Most of the reports of the formation of $OH^\cdot$ or a similar potent oxidant by PMNs are based on the detection of a specific product formed by oxidative attack on an organic precursor. The inference that the oxidant(s) is $OH^\cdot$ usually rests on the inability of either O_2^- or H_2O_2 alone to oxidize the organic precursor while in combination the precursor is oxidized. Ancillary evidence for the formation of $OH^\cdot$ is comprised of inhibiting the formation of the oxidized product by compounds known to react avidly with $OH^\cdot$. The formation of ethylene as the product of the oxidation both of methional [127–131] and of 2-keto-4-thio methylbutyric acid (KMB) [128,130–132] in the presence of activated PMNs has been measured by gas chromatography. Methane and ethane have been determined as the products of the oxidation of dimethyl sulfoxide [133] and the formation of methionine sulfone was assayed from methionine [134].

The mechanism of the formation of oxidizing species as determined by the effects of specific inhibitors have varied. In suspensions of both phagocytosing PMNs [127] and of granules derived therefrom [128] ethylene was produced from methional or from KMB. In both cases superoxide dismutase inhibited formation of ethylene but catalase did not. Nor did H_2O_2 stimulate the formation of ethylene in the suspensions of granules [128], leading the authors to propose that H_2O_2 was not an obligatory precursor of the oxidant. In other studies of PMNs stimulated with opsonized zymosan [129,130] catalase did inhibit formation of ethylene as did superoxide dismutase, ethanol, benzoate. The lack of inhibition by dimethylfuran was thought to mean that $^1O_2$ was not involved [130]. Inhibition by azide and cyanide was not explained.

The inhibitory effects of azide and cyanide might be explained by the findings of Klebanoff and Rosen [131] who, after eliminating the formation of ethylene which was not dependent on the presence of cells, observed formation of ethylene apparently dependent on myeloperoxidase. Formation of ethylene was diminished in PMNs from patients with myeloperoxidase deficiency. Myeloperoxidase restored the formation of ethylene by these cells to normal. A cell-free system consisting of myeloperoxidase, H_2O_2, Cl^-, and EDTA caused the formation of ethylene from KMB. In this cell-free system all the components were necessary, Fe^{3+} stimulated formation of ethylene, and native but not heated superoxide dismutase inhibited. The last observation was particularly puzzling since no source of O_2^- was present. Similar findings were reported by Takanaka and O'Brien [132] who also found that the sequence in which products of the burst were detected in activated PMNs was first the formation of O_2^- followed by chemiluminescence followed by formation of ethylene from KMB. Hydrogen peroxide, O_2^-, and $OH^\cdot$ were all judged to be involved in the formation of methane from dimethyl sulfoxide in the presence of stimulated PMNs [133]. In none of these cases can the identity of the potent oxidant(s) be stated with certainty.

As a means of identifying the potent oxidant another approach was made. By the use of certain organic compounds one can infer the presence of a free radical by converting a highly unstable one such as $OH^\cdot$ to a much more stable compound which can accumulate in solution, rendering its detection by electron spin resonance possible. To measure the presence of $OH^\cdot$ the compound DMPO (5,5-dimethyl 1-pyrroline-N-oxide) has been used [135,136]. The characteristic adduct of $OH^\cdot$ with DMPO was detected in suspensions of PMNs which had been stimulated with opsonized zymosan [135,136]. The formation of the signal was inhibited by superoxide dismutase and by mannitol but not by catalase [135,136]. Polymorphonuclear neutrophils from a patient with myeloperoxidase deficiency formed greater amounts of the adduct than did normal PMNs. These data, in striking contrast to the findings in some of the experiments measuring the formation of the potent oxidant (see above, reference [131]) were proposed to be explained by the formation of $OH^\cdot$ by a reaction which did not involve H_2O_2 [135].

$$Fe^{2+} + HO_2^{\cdot} \rightarrow Fe^{3+} - O_2H^- \quad (10)$$

$$Fe^{3+} - O_2H^- + e^- + H^+ \rightarrow Fe^{3+} + OH^- + OH^{\cdot} . \quad (11)$$

The data with the spin trap DMPO were very similar to those of Tauber et al. [127,128] measuring the formation of ethylene from methional or KMB. Unfortunately, as with the use of oxidizable organic molecules, the specificity for $OH^{\cdot}$ of the spin adduct identified in the experiments with ESR is not clear. A critical, cautionary review of this approach has recently appeared [137].

Still another highly reactive molecule which is suspected of being formed as a consequence of the respiratory burst is $^1O_2$. The supposition that $^1O_2$ was a product of the respiratory burst was initially based on the chemiluminescence of phagocytosing PMNs [138,139] but this interpretation was questioned on the basis of the spectrum of emission of the chemiluminescent species [140]. This objection is somewhat vitiated by the observation that biological compounds like bovine serum albumin and tryptophan shift the spectrum of the light emitted by $^1O_2$ to shorter wavelengths [141] rendering its spectral identification difficult. The formation of $^1O_2$ has also been inferred from the increased susceptibility of a mutant of *Sarcina lutea* which lacked carotenoids to killing by PMNs [142]. Since carotenoids quench $^1O_2$ the resistance of the carotenoid-containing parent strain was attributed to its greater ability to prevent the injurious effects of $^1O_2$. Finally, Rosen and Klebanoff noted the conversion, by myeloperoxidase, H_2O_2, and halide, of 2,3-diphenylfuran to cis-1,2-dibenzoylethylene, a reaction thought to be specific for $^1O_2$. Because known quenchers of $^1O_2$ inhibited the reaction and D_2O, which increases the lifespan of $^1O_2$, stimulated the reaction, $^1O_2$ was proposed to have been formed [143]. However, this reaction has subsequently been shown not to be specific for $^1O_2$ [144,145]. Nevertheless, the effects of the quenchers and of D_2O remain compatible with the inference that $^1O_2$ is formed.

The roles of O_2^- and its products in the microbicidal function of PMNs. The intimate association of the respiratory burst with the act of engulfing microorganisms and the impaired respiratory burst in PMNs from patients with chronic granulomatous disease in whom the killing of certain bacteria is defective prompted experiments to assess the roles of O_2^- and its progeny in the killing of bacteria. The results have been mixed. Yost and Fridovich [146] approached the problem by assessing the killing by PMNs of *E. Coli* in which the content of superoxide dismutase had been manipulated experimentally. By varying the conditions in which the bacteria were grown (aeration, Fe content of the medium) organisms were obtained which contained either lesser concentrations of the MnSOD but normal amounts of the FeSOD or greater concentrations of the Mn-superoxide dismutase and lesser concentrations of the Fe-Superoxide dismutase. The organisms with a reduced content of the Fe-superoxide dismutase were killed in whole blood after a shorter lag and more rapidly than the cells in which the Mn superoxide dismutase was diminished. To correct for the effects of growth in Fe-deficient medium on catalase and peroxidase ($^2/_3$ and $^1/_{12}$ respectively of control activities), the susceptibility to killing of bacteria grown either in Fe-rich or in Fe-depleted medium was compared in the presence of cyanide to inhibit the bacterial heme-containing enzymes peroxidase and catalase. The cells grown in Fe-depleted medium still were more susceptible to killing, suggesting that even when the effects of catalase and peroxidase were eliminated with CN^-, an iron-dependent activity was required for resistance. This activity was inferred to be the Fe-superoxide dismutase. This difference in rate of killing was thought not to be due to differences in rates of phagocytosis because the killing under anaerobic conditions, when phagocytosis is not impaired, was actually greater for the iron-replete than for the iron-deficient organisms. The experiments with CN^- are complicated by the inhibition by CN^- of myeloperoxidase, a potent microbicidal enzyme in PMNs. Finally, *E. Coli* with differing contents of the Mn- and Fe-superoxide dismutases were subjected to killing by a photochemical flux of O_2^-, produced by riboflavin-light and methionine. The rate of killing was greatest in bacteria with the least content of the Fe-superoxide dismutase and least in bacteria with the greatest content. Data were not presented, however, showing that the killing in the photochemical system was due to O_2^-; the killing of bacteria in blood could not be inhibited with superoxide dismutase, a finding interpreted to mean tha the enzyme was excluded from the phagosome.

Johnston et al. [123] have provided extensive experimental evidence for the importance of O_2^- (and H_2O_2) in the bactericidal repertoire of PMNs. In contrast with the results of Yost and Fridovich superoxide dismutase inhibited the killing of three pathogenic bacteria by PMNs as did catalase. The inhibitory effects of catalase and superoxide dismutase were much greater when they had adhered to latex particles, allowing them to be enclosed with the bacteria within the phagocytic vacuole. Their greater effects when so attached were attributed to a more advantageous location wherein they could scavenge O_2^- or H_2O_2 before either could react with bacteria with which they shared the phagosome. The data are slightly complicated by the effects of the latex particles themselves which, even without enzymes attached, reduced the killing of bacteria and which are known to stimulate the burst. The inhibitory effects on bacterial killing of benzoate and mannitol were felt to reflect their abilities to react with $OH^·$. The inhibitory effects of superoxide dismutase on bacterial killing by PMNs has been confirmed [147].

The bactericidal effects of O_2^- or its progeny were further supported by experiments in a cell-free system in which O_2^- was generated by the enzymic action of xanthine oxidase on acetaldehyde [148]. The killing of *Staphylococcus aureus* was inhibited by superoxide dismutase, catalase, mannitol, benzoate, azide, histidine, diazabicyclo-octane, xanthine, urate, and hypoxanthine. The effects of superoxide dismutase and catalase were attributed to their effects on their substrates, O_2^- and H_2O_2, respectively and the effects of the other compounds to their abilities to scavenge $OH^·$ and/or $^1O_2$. The possibility that $^1O_2$ was one of the reactive, bactericidal species was suggested by the greater susceptibility to killing of the mutant of *Sarcina lutea* which lacks carotenoids [142]. The concentration of substrate for xanthine oxidase could be reduced one-hundred fold in the presence of myeloperoxidase and a halide to achieve the same extent of killing, indicating far more efficient use of the products of the enzymic reaction when myeloperoxidase was present. In the presence of myeloperoxidase superoxide dismutase no longer inhibited, suggesting that O_2^- was no longer required. The necessity of H_2O_2 was shown by the inhibitory effects of catalase. Azide may have acted either by inhibiting myeloperoxidase or by scavenging $^1O_2$. The effects of 1,4-diazobicyclo(2,2,2)octane and histidine might be attributed to their scavenging of $^1O_2$ but distinguishing between an effect on $OH^·$ and an effect of

$^1O_2$ is difficult. These experiments demonstrate the great bactericidal efficiency of myeloperoxidase.

A dissenting view which was based partly an another approach has been published [149]. The roles of H_2O_2 and O_2^- in the killing of *Staph. aureus* were assessed by correlating the bacterial content of catalase and of superoxide dismutase with the pathogenecity of the organism. Using isolates of *Staph. aureus* from a clinical laboratory, a correlation between the activity of catalase in *Staph. aureus* and lethality in mice was observed ($r = 0.88$) but no correlation was found between lethality and the content of superoxide dismutase. Catalase could inhibit the killing of an organism in which the activity of catalase was decreased but superoxide dismutase could not. Similarly, injection of catalase intraperitoneally increased the virulence of *Staph. aureus* in mice but the injection of superoxide dismutase did not. Certainly the lack of any effect of the superoxide dismutase within the bacterium on O_2^- produced outside could be explained by the barrier to the entry of O_2^- posed by the cell wall, for which there is experimental support [150]. The absence of an effect on the virulence of staph of injected superoxide dismutase could indicate that O_2^- is not an important microbicidal species but other explanations can also be advanced. Even when superoxide dismutase is present in the medium surrounding phagocytosing PMNs its effect on killing of bacteria is modest unless it is forced into the phagosome by attachment to an ingestable particule [123] or by altering its charge by coupling to polylysine thereby increasing its positive charge to allow its approach to the negatively charged cell membrane [151]. Given these qualifications and the very short survival in the circulation of unmodified CuZn superoxide dismutase from bovine erythrocytes [152,153], the concentration of superoxide dismutase may have been so low that an inhibitory effect on bacterial killing could not have been observed.

The killing of other types of organisms by PMNs has also been assessed. Klebanoff [154] observed that. *E. Coli*, *Staph. aureus*, and *Candida tropicalis* were all killed in a cell-free system in which O_2^- and H_2O_2 were both generated by the enzymic action of xanthine oxidase on xanthine in the presence of myeloperoxidase and Cl^-. In this system, O_2^- appeared to play no role as shown by the lack of inhibition by superoxide dismutase except at very high concentrations at which the denatured enzyme also inhibited killing. All components of the system were shown to be necessary (xanthine oxidase, xanthine, myeloperoxidase, and Cl^-). The great dependency on H_2O_2 as a substrate for myeloperoxidase was shown by an experiment in which killing was inhibited by cytochrome c, which by reacting with the O_2^- and returning it to solution as dioxygen, could diminish the formation of H_2O_2. That this was its mode of action was shown by reversing the inhibition produced by cytochrome c with superoxide dismutase which competes with cytochrome c for O_2^- and causes the formation of H_2O_2.

The larvae of *Trichinella spiralis*, the parasitic cause of trichiniasis, are also susceptible to killing by products of the respiratory burst. In this case H_2O_2 itself appears to be the active species. In the presence of immune serum PMNs killed larvae of *T. spiralis* but PMNs from patients with chronic granulomatous disease in which the burst is markedly impaired killed poorly [155]. Catalase inhibited the killing but superoxide dismutase did not. Azide and cyanide which inhibit myeloperoxidase enhanced killing. These data which suggested that H_2O_2 might itself be cytotoxic for the larvae of *T. spiralis* were confirmed in studies in a cell-free system [156]. When both O_2^- and H_2O_2 were produced by the enzymic action of xanthine oxidase on xanthine, larvae were killed. Killing was inhibited by catalase but not by superoxide dismutase. The enzymic action of glucose oxidase on glucose which forms H_2O_2 but not O_2^- was also larvicidal. Suspensions of particles from a homogenate of either PMNs or eosinophils, which contained myeloperoxidase, inhibited killing, indicating that H_2O_2 may have been directly toxic to these larvae. Myeloperoxidase, which uses H_2O_2 as a substrate to form its bactericidal products, presumably acted by diminishing the concentration of H_2O_2. Indeed, larvae of *T. spiralis* were shown to be markedly more sensitive to H_2O_2 than bacteria.

The killing of both *Candida albicans* [157)] and *Schistosoma mansoni* [158)] by PMNs appeared to be dependent on myeloperoxidase. The PMNs from patients either with deficiency of myeloperoxidase or with chronic granulomatous disease killed *C. albicans* poorly [157)]. A cell-free system of myeloperoxidase, a halide, and a source of H_2O_2 killed efficiently. All the components of the system were necessary. Either catalase or inhibitors of myeloperoxidase inhibited the killing as did quenchers of $^1O_2$. Similarly *Schistosoma mansoni* stimulated the formation of H_2O_2 by PMNs and was killed in a manner which was inhibited by catalase and by inhibitors of myeloperoxidase, CN^- and N_3^- [158)].

2.8.1 Other Cytocidal Effects

In certain cases the killing of mammalian cells had been shown to result from products of the respiratory burst in PMNs. The longevity of PMNs themselves as assessed by their ability to exclude the supravital dye, trypan blue, was increased during phagocytosis by superoxide dismutase [159)]. The toxicity of H_2O_2 or its products was studied in the PMNs from subjects with the inherited reduction in the activity of catalase [160)]. In the PMNs from affected individuals (with about 25% of the normal activity) which were exposed to a source of H_2O_2 generated continuously from glucose and glucose oxidase, the ability to undergo the respiratory burst in response to zymosan fell. The function of PMNs as assessed by their ability to engulf droplets of paraffin oil and to migrate toward a chemotaxin was increased by catalase but not by superoxide dismutase [161)]. When H_2O_2 and O_2^- were generated externally, again catalase but not superoxide dismutase preserved normal uptake. That the toxicity of H_2O_2 might be exerted in part through the enzymic action of myeloperoxidase was suggested by the demonstration that the combination of myeloperoxidase, H_2O_2, and a halide produced lysis as demonstrated by the release or ^{51}Cr from PMNs [162)].

Erythrocytes, too, are susceptible to the reactive products of the respiratory burst. In one study, erythrocytes from patients with the inherited lack of glucose-6-phosphate dehydrogenase, an enzyme in the hexose monophosphate shunt which is necessary for the formation of NADPH, were the targets. Because NADPH is the necessary source of electrons for the reduction of glutathione in the erythrocyte, these cells are unable to maintain enough reduced glutathione for the reaction catalyzed by glutathione peroxidase which detoxifies H_2O_2. These cells are, therefore, markedly vulnerable to H_2O_2. When PMNs were exposed to latex particles to stimulate the burst, the erythrocytes deficient in glucose-6-phosphate dehydrogenase were so injured that when reinfused into the subject they were removed from blood much more rapidly than were normal erythrocytes treated similarly [163)].

Normal erythrocytes have also been lysed by PMNs activated with phorbol myristate acetate [164,165]. Both superoxide dismutase and catalase inhibited the lysis but scavengers of $OH^{\cdot}$ and $^1O_2$ did not [163]. The mechanism of this effect was dissected by treating the erythrocytes with CO which prevented lysis. The effect of CO, which binds avidly to hemoglobin, suggested that lysis required the participation of hemoglobin. Since O_2^- is known to cause the formation of methemoglobin from oxyhemoglobin [201,207–209], this was proposed as the effect of O_2^-. In accord with this proposal, when methemglobin was formed first by treating the erythrocytes with nitrite, and the red cells were then subjected to the activated PMNs, the inhibitory effects of superoxide dismutase were lost. The lysis of erythrocytes containing methemoglobin was markedly inhibited by catalase. An explanation was proposed in which O_2^- first caused the formation of methemoglobin which could then react with H_2O_2 to form an oxidant which attacked the erythrocyte from within [165]. A prediction based on this formulation has subsequently been tested; superoxide appears to enter the red cell through the channel in the membrane for the exchange of stable anions [215]. By blockade of this channel the effects of O_2^- on the formation of methemoglobin should be abolished. Indeed, this was observed [166]. Certainly these experiments suggest one way in which O_2^- formed by PMNs can damage cells. The unique abundance of hemoglobin in erythrocytes, however, may limit the relevance of this mechanism to other cells.

Lysis of antibody-coated red cells through the products of the respiratory burst appears to be linked to phagocytosis [167]. Neutrophils from patients with chronic granulomatous disease lysed antibody-coated erythrocytes less effectively than PMNs from normal subjects in suspension, but when phagocytosis was prevented by attachment of the erythrocytes to plastic or by treatment with colchicine, PMNs from the patients with chronic granulomatous disease lysed the red cells normally.

Neutrophils activated with zymosan-activated serum damaged cultured endothelial cells as shown by release of ^{51}Cr in a manner which was inhibited by catalase and by the combination of catalase and superoxide dismutase but not by superoxide dismutase alone [168].

The antibody-dependent lysis of tumor cells by PMNs exhibited some of the characteristics of damage mediated by products of the burst; in the presence of tumor cells there was increased consumption of O_2, increased formation of O_2^- and activation of the hexose monosphosphate shunt [169]. However, although a reduction in the concentration of O_2 in the medium inhibited lysis neither catalase nor superoxide dismutase inhibited. The lack of effect on these enzymes was attributed to their inability to interpose themselves between the plasma membranes of the PMN and its target. Similar conclusions were reached by Clark, and Klebanoff [170] whose data incriminated the products of the burst by the reduced killing of tumor cells by PMNs from patients with chronic granulomatous disease. Myeloperoxidase, however, appeared not to be required since neither azide or cyanide inhibited and killing by PMNs from patients with inherited deficiency of myeloperoxidase was normal.

However, when PMNs were stimulated, not by antibody on the surface of the tumor cell, but instead by Concanavalin A [171] or by opsonized zymosan [172], myeloperoxidase did appear to mediate the killing; azide and cyanide inhibited the killing, halides were required, catalase inhibited, and PMNs from patients with either hereditary deficiency of myeloperoxidase or chronic granulomatous disease were ineffective. Similar effects of azide and cyanide on the lysis of lymphocytes by PMNs after stimulation with phorbol myristate acetate have been reported [173] and an impressive lytic effect of HOCl, the product of the myeloperoxidase reaction, was demonstrated. Myeloperoxidase may not be involved in the antibody-dependent killing of tumor cells if release of the contents of the lysosomes where myeloperoxidase is stored did not occur.

Still a different set of results was obtained by Nathan et al. [174, 175] who assessed the tumoricidal effect of macrophages or of PMNs stimulate by PMA. These authors concluded that H_2O_2 itself was the lytic species. Catalase but not superoxide dismutase inhibited lysis. The curves relating the dose of PMA either to the formation of H_2O_2 by PMNs or to killing of tumor cells by PMNs were identical. Furthermore, there was a linear relationship between the susceptibility of each type of tumor to lysis by H_2O_2 and to its susceptibility to lysis by PMNs. Lactoperoxidase and I^- inhibited rather than stimulated cytotoxicity in the presence of serum.

The H_2O_2 which PMNs elaborate during the burst affects platelets as well. Hydrogen peroxide could diffuse from phagocytosing PMNs across a dialysis membrane and cause inhibition of aggregation of platelets on the opposite side of the membrane [176]. Platelets also release serotonin and adenine in the presence of PMNs stimulated with opsonized zymosan [177]. The involvement of products of the respiratory burst was shown by the diminished effects of PMNs from patients with either hereditary deficiency of myeloperoxidase or with chronic granulomatous disease on platelets. The effects on platelets were blocked by catalase or by omission of halides, further suggesting that myeloperoxidase mediated these effects, but H_2O_2 itself in the absence of myeloperoxidase had similar effects. The authors suggested that these effects of H_2O_2 derived from the respiratory burst could be mediated through the enzymic action of myeloperoxidase or by H_2O_2 itself.

Chemiluminescence. That PMNs stimulated to undergo the respiratory burst cause the emission of light has already been mentioned. Whatever the species which actually emit photons, the phenomenon appears to be dependent on O_2^- and H_2O_2 [178, 179], seems faithfully to reflect the activity of the burst [180–182], is enhanced by scavengers of $OH^\cdot$ [183], and is amplified by the cyclic hydrazide, luminol [181, 184]. The simplicity with which chemiluminescence is measured in a liquid scintillation counter with the coincidence circuits off has made it popular with students of the respiratory burst in PMNs.

2.8.2 The Role of Products of the Burst in Inflammation

The ability of PMNs to elaborate O_2^- and its reactive progeny and the capacity of these progeny to react with many components of cells attracted the interest of investigators of the molecular basis of the inflammatory response. One way in which PMNs may participate is by acting as the source of the molecules such as $OH^\cdot$ which may destroy tissue. The inhibition of the inflammatory effects of Carrageenan by superoxide dismutase [185] might represent such an effect.

A second way in which PMNs may affect the magnitude of the inflammatory response is by regulating their own movement to sites of inflammation. Evidence that this may be a function of which PMNs are capable has been provided by Petrone et al. [153] who attempted to modify the inflammation elicited by the reverse passive

Arthus reaction with superoxide dismutase. To ensure that the enzymes which were to be injected into animals were likely to survive long enough to have an effect, superoxide dismutase was covalently coupled to large polysaccharides to increase the molecular weight of the enzyme so that it would not be filtered at the glomerulus and thereby lost into the urine. Animals were injected intravenously with antigen (human serum albumin) and intradermally with antibody. Inflammation at the site of the injection of the antibody was completely suppressed by an intravenous injection of Ficoll-superoxide dismutase but not by Ficoll or by catalase. The migration of PMNs to the site of the inflammatory response as assessed histologically was drastically curtailed. Experiments in vitro demonstrated that a chemotactic material was formed from a lipid-like precursor in plasma by the action of xanthine oxidase on xanthine, an enzymic means of generating O_2^-. Its formation was inhibited by superoxide dismutase but not by catalase, suggesting that when PMNs are activated in the presence of plasma a chemotactic material may be formed from O_2^-. Perez et al.[186] have demonstrated the formation of a chemotactic factor from arachidonic acid in the presence of an enzymic source of O_2^-. The formation of the chemotatic material was inhibited both by scavengers of O_2^- and of H_2O_2 and by scavengers of $OH^{\cdot}$. Whether the chemotactic materials of Petrone et al. and of Goldstein et al. are identical and whether they are formed in vivo as a means of regulating the migration of PMNs is not clear. Clearly the patterns of inhibition were different, at least implying that the mechanisms of forming these two chemotaxins differ.

Some of the effects of products of the respiratory burst would appear to have opposing effects on the inflammatory response. Neutrophils stimulated with an antigen-antibody complex or by an enzymic source of O_2^- inactivated the elastase-inhibitory activity in serum[187,188]. Inactivation was prevented by both catalase and superoxide dismutase and by scavengers of $OH^{\cdot}$, suggesting a reaction of the Haber-Weiss type (reaction 4). However, involvement of myeloperoxidase in PMNs was also suggested by inhibiting the inactivation with azide, cyanide, and depletion of Cl^- and by reproducing the inactivation by myeloperoxidase, H_2O_2, and Cl^-. Clark et al.[189] have recently demonstrated the ability of myeloperoxidase, in conjunction with H_2O_2 and Cl^-, to inactivate the alpha$_1$ protease inhibitor of serum. Similar effects were produced by PMNs activated with phorbol myristate acetate. The inhibition by CN^-, N_3^- and catalase as well as the lack of inactivation by PMNs from patients with deficiency of myeloperoxidase suggested that the myeloperoxidase in PMNs catalyzed the destruction of the alpha$_1$ antiprotease. One would suppose that the effect of inhibition of the inactivator of elastase as a consequence of the respiratory burst would be to increase the damaging effects of unopposed elastolysis.

Contrary to this anticipated effect is the effect of products of the burst on chemotactic factors. Clark and Klebanoff[190] showed that both chemotaxins, C_{5a} and FMLP, lost chemotactic activity in the myeloperoxidase system. The anticipated consequence of this action of myeloperoxidase would be to terminate the influx of PMNs into an inflamed site.

Another means by which PMNs may help terminate the inflammatory consequences of their activation is by the secretion of lysozyme from their lysosomal granules. Treatment of PMNs with lysozyme not only reduced the chemotaxis of PMNs in vivo and in vitro but also diminished the formation of O_2^- in response to opsonized zymosan and to phorbol myristate acetate [191]. That this effect of lysozyme was the consequence of its enzymic action was demonstrated by partial reversal of these effects by a specific inhibitor, tri-N-acetyl glucosamine.

2.9 Chronic Granulomatous Disease

The relative rarity of the inherited illness, chronic granulomatous disease, does not indicate its great importance in the evolution of the understanding of the respiratory burst in PMNs. In this disease (reviewed in reference 95) children are afflicted with severe, recurring, infections with bacterial pathogens, but only with certain organisms. The common feature of the bacteria which do cause infection is that they do not excrete H_2O_2, usually as a consequence of the synthesis by the bacterium of catalase [192]. The abnormalities in the function of PMNs all appear attributable to a defect in the respiratory burst. After stimulation of PMNs there is reduced consumption of O_2 [7], diminished formation of O_2^- [123,193–195] and H_2O_2 [7,125,104], defective activation of the hexose monophosphate shunt [7], lessened chemiluminescence [123], and impaired killing of those bacteria [7] which require products of the burst to be killed. Phagocytosis is normal. The activity of the NADPH oxidase is reduced in stimulated PMNs [99,100,196]. That the defect may be in the mechanism by which the burst is initiated has been inferred from the defective change in polarization of the membrane in response to appropriate stimuli [60,61] as described under "Signal Transmission."

Although the usual patient is a boy, as expected with an illness inherited as an X-linked recessive, a substantial minority of patients are girls [95]. Recently this has been proposed to result from random X-inactivation because the PMNs from mothers of afflicted girls were found to exhibit reduced chemiluminescence in response to opsonized zymosan [197]. The chemiluminescence in the mothers was intermediate between normal and the severely diminished levels in the afflicted children. In each pedigree the PMNs of either the patients or their mothers were composed of two populations of cells with respect to the reduction of nitroblue tetrazolium; one population reduced nitroblue tetrazolium normally and the other did not. This finding is consistent with the existence of one population of PMNs in which the mutant chromosome was inactivated, allowing normal reduction of nitroblue tetrazolium by O_2^-, and a second population in which the normal chromosome was inactivated leaving only the mutant X chromosome to code for a gene product(s) which was defective in a necessary component of the burst. In the pedigrees of afflicted females only on the maternal side were there afflicted males, parents with intermediate chemiluminescence by PMNs, or female siblings with diminished chemiluminescence by PMNs.

Among the questions which remain unanswered in this illness is the function of O_2^-. The defect in killing of bacteria appears correctable with a means of supplying H_2O_2 to the interior of the phagosome. The bacterium may do so if it lacks catalase and consequently excretes H_2O_2 [192]. Another particle such as a latex bead to which an enzymic source of H_2O_2 is attached [198,199] will also suffice. In either of these cases killing of bacteria is improved. If H_2O_2 corrects the defect what, then, is the necessity for the formation of O_2^- at all? The evolution of an enzyme which forms O_2^- solely for the purpose of manufacturing H_2O_2 seems unnecessarily cumbersome and potentially injurious since many other flavoenzymes form H_2O_2

directly without any O_2^- as precursor. Are there other roles of O_2^- in PMNs? Are there other defects in these patients for which the lack of the production of O_2^- accounts?

3 Erythrocytes

The non-nucleated erythrocyte of mammals is rich in the CuZn superoxide dismutase. In the mature cell the nucleus, ribosomes, mitochondria have all disappeared leaving a plasma membrane enclosing a concentrated solution of hemoglobin and those enzymes required for the preservation of its function in the transport of oxygen and CO_2. It is tempting to suppose that in these marvels of specialization enzymes which have been retained serve some vital function. The richness of the erythrocyte in superoxide dismutase, in this, light, might appear mysterious because the means by which erythrocytes form O_2^- is not obvious. The reactions in which O_2^- is formed represent reductions of molecular oxygen, reactions which occur chiefly in organelles which the erythrocytes lack. Obviously, the erythrocyte functions in environments where oxygen abounds and transports oxygen in its circulatory rounds, but it does not normally reduce the oxygen it carries. What, then, is the source of O_2^- in the erythrocyte which has made it advantageous to contain large amounts of superoxide dismutase? What is the effect of O_2^- on the erythrocyte from which superoxide dismutase protects it?

Certain answers to these questions can be offered. The most persuasive answer to the question of the source of O_2^- is that oxyhemoglobin itself is the culprit. Although dioxygen is usually released from oxyhemoglobin, some of the time oxyhemoglobin may dissociate to ferrihemoglobin (methemoglobin) and O_2^-:

$$\underset{\text{(oxyhemoglobin)}}{HbFe^{2+} - O_2} \rightarrow \underset{\text{(methemoglobin)}}{HbFe^{3+}} + O_2^- . \tag{12}$$

The 'pathologic' dissociation has been demonstrated in the hemoglobin of the shark, a creature whose hemoglobin appears to be particularly prone to this dissociation [200], in human oxyhemoglobin [201], and in the heterodimers (alpha$_1$beta$_1$) of hemoglobin [202]. Superoxide is also produced when oxyhemoglobin is illuminated [203] and when hemoglobin reacts with certain organic compounds [204,205]. Whether any of these reactions form O_2^- in sufficient amounts to form the pressure that has caused the non-nucleated red cell to retain such an abundance of superoxide dismutase is still not clear.

Because the erythrocyte is a relatively simplified cell with a circumscribed function it would seem that superoxide dismutase must act to prevent a deleterious effect of O_2^- on the erythrocyte's capacity to transport oxygen. Superoxide might react with hemoglobin to form a derivative which cannot reversibly bind O_2^-. One such reaction is to mediate the formation of methemoglobin, the derivative of hemoglobin in which the Fe^{2+} in heme is converted to Fe^{3+}. The ability of O_2^- to react with hemoglobin to form methemoglobin has been proposed to occur [201,207-209] in a reaction in which the bound O_2 is reduced by O_2^-, permitting the subsequent dissociation of methemoglobin and peroxide:

$$HbFe^{2+}{-}O_2 + O_2^- + 2\,H^+ \rightarrow HbFe^{3+} + H_2O_2 + O_2\,. \qquad (13)$$

The evidence for this reaction is that when solutions of hemoglobin are subjected to a flux of superoxide anions generated either photochemically [201, 207–209] or by radiolysis [207] the formation of methemoglobin occurs as determined by the characteristic change in absorbance. The formation of methemoglobin could be inhibited by superoxide dismutase but not by its catalytically inactive derivative. Whether superoxide reacts with hemoglobin in red cells to form methemoglobin in vivo is not clear. It seems unlikely that this occurs because the rate of the reaction is relatively slow.

Another possible effect of O_2^- on erythrocytes is to initiate injury to the cell membrane. Such a lesion might shorten the lifespan of the cell. Several studies have suggested that O_2^- can indeed disrupt the membrane of the human red cell, probably indirectly. Intact erythrocytes have been subjected to a flux of O_2^- generated externally by the enzymic action of xanthine oxidase [210] and resealed ghosts from erythrocytes have been subjected to a flux of O_2^- generated internally [211]. In both cases there was inhibition of lysis by superoxide dismutase but not by the heated enzyme. Catalase inhibited the lysis of the intact cells but not of the resealed ghosts. A variety of other scavengers also inhibited lysis, but the lack of specificity of these prevented unequivocal interpretation of the data. It seemed likely that some reactive progeny of O_2^- was the species but whether this was $OH^{\cdot}$ or $^1O_2$ or some other molecule is not clear. In the case of the intact cells a reaction between H_2O_2 and O_2^- was proposed.

$$H_2O_2 + O_2^- \rightarrow OH^{\cdot} + OH^- + {}^1O_2\,. \qquad (14)$$

The lack of inhibition of lysis by catalase in the resealed vesicles prompted the proposal that the precursor which reacted with O_2^- to form the lytic species was not H_2O_2 but instead a lipid hydroperoxide; the marked sensitization of the vesicles to the lytic effects of O_2^-, which prior photooxidation produced, was consistent with this proposal [211]. Lysis of bovine red cells exposed to O_2^-, however, was inhibited by superoxide dismutase but not by catalase [212].

Clarification of the differing inhibitory effects of catalase in the intact human erythrocytes and in the resealed vesicles may be provided by the experiments of Weiss and colleagues [165] whose data suggest that in the intact erythrocyte subjected to a flux of O_2^- from PMNs there are at least two steps leading to lysis. In the first O_2^- permeated the cell membrane and reacted with oxyhemoglobin to yield methemoglobin. Once methemoglobin was formed there was no longer inhibition of lysis by superoxide dismutase. Methemoglobin could be formed in the erythrocyte by other means and when these cells were exposed to PMNs stimulated with PMA, lysis was inhibited by catalase but not by superoxide dismutase. It was proposed that methemoglobin reacted with H_2O_2 within the erythrocyte to form a powerful oxidant which attacked the membrane from within. These observations predict that both catalase and superoxide dismutase would inhibit the lysis of intact erythrocytes in a flux of O_2^- but that in vesicles, from which hemoglobin had been removed by washing before resealing, some other mechanism must prevail.

Experiments in which the lysis of erythrocytes exposed to oxidizable organic compounds have also been performed [204–206, 213, 214]. In some of these experiments O_2^- and H_2O_2 appear to be involved as shown by the inhibition of lysis by superoxide dismutase and catalase. However, the pathways leading to lysis in these experiments may be complex because of the effects of O_2^- and H_2O_2 on the rates of autoxidation of the organic compounds [210] and because of the possibility that the oxidizable organic compounds or their products react with the red cells directly.

The ability of O_2^- to convert oxyhemoglobin to methemoglobin and to initiate disruption of the membrane of the erythrocyte may be effects that superoxide dismutase acts in the erythrocyte to prevent. There is, however, another possibility. Superoxide dismutase may exist in red cells to scavenge O_2^- formed in plasma. The results of experiments consist with this possibility have been presented [215]. In these experiments xanthine oxidase was sealed within vesicles prepared from the membranes of erythrocytes. When these vesicles were incubated in phosphate buffer containing acetaldehyde as a substrate for xanthine oxidase, cytochrome c in the buffer was reduced. That O_2^- was the reductant was demonstrated by inhibition of the reduction of cytochrome c by superoxide dismutase added to the medium. Because the membrane of the erythrocyte contains a channel for the exchange of stable anions, it was proposed that O_2^- might move through the membrane of the red cell in this channel. Indeed, blockade of the anion channel with either of two sulfonated stilbene derivatives markedly inhibited the reduction of cytochrome c in the medium by O_2^- formed inside the vesicles. Inhibition by the blockers of the anion channel was overcome by disrupting the membrane of the vesicle with Triton X 100, suggesting that the effects of the sulfonated stilbenes were exerted only in the presence of an intact membrane. If O_2^- could indeed pass through the membrane of the red cell in the anion channel might it do so in vivo? The answer to this question is not known. However, there are some attractive features to such a proposal. Superoxide in plasma could enter the cell readily through the anion channel where it encounters superoxide dismutase which converts two superoxide anions to water and H_2O_2. The hydrogen peroxide could then be catalytically scavenged, either by catalase to water and dioxygen, or by glutathione peroxidase to water and oxidized glutathione. By this means O_2^- formed in plasma, by phagocytic PMNs for example, could be efficiently converted to water and dioxygen.

4 References

1. McCord, J. M., Fridovich, I.: J. Biol. Chem. *244*, 6049 (1969)
2. Fridovich, I.: Science *201*, 875 (1978)
3. Babior, B. M., Kipnes, R. S., Curnutte, J. T.: J. Clin. Invest. *52*, 741 (1973)
4. Baldridge, C. W., Gerard, R. W.: Amer. J. Physiol. *103*, 235 (1933)
5. Iyer, G. Y. N., Islam, D. M. F., Quastel, J. H.: Nature *192*, 535 (1961)
6. Selvaraj, R. J. Sbarra, A. J.: Nature *211*, 1272 (1966)
7. Holmes, B., Page, A. R., Good, R. A.: J. Clin. Invest. *46*, 1422 (1967)
8. Quie, P. G., et al.: J. Clin. Invest. *46*, 668 (1967)
9. Klebanoff, S. J., Clark, R. A.: The Neutrophil: Function and Clinical Disorders, North Holland Amsterdam, 1978
10. Haber, F., Weiss, J.: Proc. Roy. Soc. (London) *A147*, 332 (1934)
11. McCord, J. M., Day, E. D., Jr.: FEBS Lett. *86*, 139 (1978)

12. Weinstein, J., Bielski, B. H. J.: J. Am. Chem. Soc. *101*, 58 (1979)
13. Fenton, H. J. H.: J. Chem. Soc. *65*, 899 (1894)
14. Matheson, I. B. C., et al.: Photochem. Photobiol. *21*, 165 (1975)
15. Newburger, P. E., Chovaniec, M. E., Cohen, H. J.: Blood *55*, 85 (1980)
16. DeChatelet, L. R., Mullikin, D., McCall, C. E.: J. Infect. Dis. *131*, 443 (1975)
17. Curnutte, J. T., Babior, B. M.: J. Clin. Invest. *53*, 1662 (1974)
18. Goldstein, I. M., et al.: J. Clin. Invest. *56*, 1155 (1975)
19. Kiyotaki, C., et al.: Immunology *35*, 613 (1978)
20. Johnston, R. B., Jr., Lehmeyer, J. E.: J. Clin. Invest. *57*, 836 (1976)
21. Goldstein, I. M., et al.: J. Immunol. *117*, 1282 (1976)
22. Newman, S. L., Johnston, R. B., Jr.: J. Immunol. *123*, 1839 (1979)
23. Hatch, G. E., Gardner, D. E., Menzel, D. B.: J. Exp. Med. *147*, 182 (1978)
24. Becker, E. L., Sigman, M., Oliver, J. M.: Am J. Pathol. *95*, 81 (1979)
25. Lehmeyer, J. E., Snyderman, R., Johnston, R. B., Jr.: Blood *54*, 35 (1979)
26. Kitagawa, S., Takaku, F., Sakamoto, S.: J. Immunol. *125*, 359 (1980)
27. Van Epps, D. E., Garcia, M. L.: J. Clin. Invest. *66*, 167 (1980)
28. Cohen, H. J., Chovaniec, M. E.: J. Clin. Invest. *61*, 1081 (1978)
29. Cohen, H. J., Chovaniec, M. E.: J. Clin. Invest. *61*, 1088 (1978)
30. Washida, N., et al.: Biochim. Biophys. Acta *631*, 371 (1980)
31. Kakinuma, K., Minakami, S.: Biochim. Biophys. Acta *538*, 50 (1978)
32. Tanswell, A. K., Lynn, W. S.: Biochem. Biophys. Res. Commun. *90*, 911 (1979)
33. Volpi, M., Naccache, P. H., Sha'afi, R. I.: Biochem. Biophys. Res. Commun. *92*, 1231 (1980)
34. Cohen, M. S., Metcalf, J. A., Root, R. K.: Blood *55*, 1003 (1980)
35. Nabi, Z. F., et al.: Exp. Cell Res. *124*, 293 (1979)
36. Okamura, N., et al.: J. Biochem. (Tokyo) *88*, 139 (1980)
37. Nakagawara, A., Minakami, S.: Biochem. Biophys. Res. Commun. *64*, 760 (1975)
38. Curnutte, J. T., Babior, B. M., Karnovsky, M. L.: J. Clin. Invest. *63*, 637 (1979)
39. Klempner, M. S., et al.: J. Clin. Invest. *64*, 996 (1979)
40. Repine, J. E., et al.: J. Lab. Clin. Med. *83*, 911 (1974)
41. DeChatelet, L. R., Shirley, P. S., Johnston, R. B., Jr.: Blood *47*, 545 (1976)
42. Simchowitz, L., Mehta, J., Spilberg, I.: J. Lab. Clin. Med. *94*, 403 (1979)
43. Kitagawa, S., Takaku, F., Sakamoto, S.: J. Clin. Invest. *65*, 74 (1980)
44. Goldstein, B. D., et al.: Biochem. Biophys. Res. Commun. *88*, 854 (1979)
45. Goldstein, I. M., et al.: J. Exp. Med. *146*, 483 (1977)
46. Goldstein, I. M., et al.: Inflammation *1*, 305 (1976)
47. Simchowitz, L., Mehta, J., Spilberg, I.: Arthritis Rheum. *22*, 755 (1979)
48. Nelson, D. H., Ruhmann-Wennhold, A.: J. Clin. Endocrinol. Metab. *46*, 702 (1978)
49. Hirata, F., et al.: Proc. Natl. Acad. Sci. USA *77*, 2533 (1980)
50. Dahnden, C., Fehr, J.: J. Clin. Invest. *66*, 884 (1980)
51. Bokoch, G. M., Reed, P. W.: Biochem. Biophys. Res. Commun. *90*, 481 (1979)
52. Smolen, J. E., Weissmann, G.: Biochem. Pharmacol. *29*, 533 (1980)
53. Oyanagui, Y.: Biochem. Pharmacol. *27*, 777 (1978)
54. Simchowitz, L., Atkinson, J. P., Spilberg, I.: J. Clin. Invest. *66*, 736 (1980)
55. Proctor, R. A.: Infect. Immun. *25*, 912 (1979)
56. Becker, E. L.: J. Reticuloendothel. Soc. *26* (Suppl.), 701 (1979)
57. Tsan, M-F., McIntyre, P. A.: J. Exp. Med. *143*, 1308 (1976)
58. Tsan, M-F., Douglass, K. H., McIntyre, P. A.: Blood *49*, 437 (1977)
59. Korchak, H. M., Weissmann, G.: Biochim. Biophys. Acta *601*, 180 (1980)
60. Seligmann, B. E., Gallin, J. I.: J. Clin. Invest. *66*, 493 (1980)
61. Whitin, J. C., et al.: J. Biol. Chem. *255*, 1874 (1980)
62. Simchowitz, L., Spilberg, I.: J. Immunol. *123*, 2428 (1979)
63. Simchowitz, L., Spilberg, I.: J. Lab. Clin. Med. *91*, 583 (1979)
64. Takeshige, K., et al.: Biochem. Biophys. Res. Commun. *95*, 410 (1980)
65. Jones, H. P., Ghai, G. M., Petrone, W. F., McCord, J. M.: Submitted for publication
66. Cohen, H. G., Chovaniec, M. E., Ellis, S. W.: Blood *56*, 23 (1980)
67. Gennaro, R., Romeo, D.: Biochem. Biophys. Res. Commun. *88*, 44 (1979)

68. Korchak, H. M., et al.: Proc. Natl. Acad. Sci. USA *77*, 2721 (1980)
69. Simchowitz, L., et al.: J. Immunol. *124*, 1482 (1980)
70. Smolen, J. E., Korchak, H. M., Weissmann, G.: J. Clin. Invest. *65*, 1077 (1980)
71. Nakagawara, A., Minakami, S.: Biochim. Biophys. Acta *584*, 143 (1979)
72. Salinhe, M. L., McCord, J. M.: J. Clin. Invest. *54*, 1005 (1974)
73. Rest, R. F., Spitznagel, J. K.: Biochem. J. *166*, 145 (1977)
74. DeChatelet, L. R., et al.: J. Clin. Invest. *53*, 1197 (1974)
75. Patriarca, P., Dri, P., Rossi, F.: FEBS Lett. *43*, 247 (1974)
76. Goldstein, I. M., et al.: J. Clin. Invest. *59*, 249 (1977)
77. Takanaka, K., O'Brien, P. J.: Arch Biochem. Biophys. *169*, 428 (1975)
78. Baehner, R. L.: J. Lab. Clin. Med. *86*, 785 (1975)
79. Dewald, B., et al.: J. Clin. Invest. *63*, 21 (1979)
80. Cohen, H. J., Chovaniec, M. E., Davies, W. A.: Blood *55*, 355 (1980)
81. Cohen, H. J., Newburger, P. E., Chovaniec, M. E.: J. Biol. Chem. *255*, 6584 (1980)
82. Briggs, R. T., et al.: J. Cell Biol. *67*, 566 (1975)
83. Iverson, D., et al.: J. Clin. Invest. *59*, 282 (1977)
84. Suzuki, Y., Lehrer, R. I.: J. Clin. Invest. *66*, 1409 (1980)
85. Kakinuma, K., Boveris, A., Chance, B.: FEBS Lett. *74*, 295 (1977)
86. Patriarca, P., et al.: Biochim. Biophys. Acta *385*, 380 (1975)
87. Takanaka, K., O'Brien, P. J.: Arch. Biochem. Biophys. *169*, 436 (1975)
88. Torres, M., Auclair, C., Hakim, J.: Biochem. Biophys. Res. Commun. *88*, 1003 (1979)
89. Auclair, C., et al.: Enzyme *23*, 225 (1978)
90. Auclair, C., Torres, M., Hakim, J.: Biochem. Biophys. Res. Commun. *81*, 1067 (1978)
91. Auclair, C., et al.: Biochim. *58*, 1359 (1976)
92. McPhail, L. C., DeChatelet, L. R., Shirley, P. S.: J. Clin. Invest. *58*, 774 (1976)
93. DeChatelet, L. R., McPhail, L. C., Shirley, P. S.: Blood *49*, 445 (1977)
94. Torres, M., Auclair, C., Hakim, J.: FEBS Lett. *97*, 50 (1979)
95. Roos, D., Weening, R. S.: Defects in the oxidative killing of microorganisms by phagocytic leukocytes, in: Oxygen Free Radicals and Tissue Damage, (ed.) Fitzsimons, D. W., pp. 225, Amsterdam, Excerpta Medica 1979
96. Badwey, J. A., Curnutte, J. T., Karnovsky, M. L.: New Engl. J. Med. *300*, 1157 (1979)
97. Selvaraj, R. J., Sbarra, A. J.: Biochim. Biophys. Acta *141*, 243 (1967)
98. Cooper, M. R., et al.: J. Clin. Invest. *51*, 769 (1972)
99. Gabig, T. G., Babior, B. M.: J. Biol. Chem. *254*, 9070 (1979)
100. Babior, B. M., Curnutte, J. T., McMurrich, B. J.: J. Clin. Invest. *58*, 989 (1976)
101. Sbarra, A. J., Karnovsky, M. L.: J. Biol. Chem. *234*, 1355 (1959)
102. Segal, A. W., Peters, T. J.: Lancet *1*, 1363 (1976)
103. Babior, B. M., Curnutte, J. T., Kipnes, R. S.: J. Clin. Invest. *56*, 1035 (1975)
104. Briggs, R. T., Karnovsky, M. L., Karnovsky, M. J.: J. Clin. Invest. *59*, 1088 (1977)
105. Root, T. K., Metcalf, J. A.: J. Clin. Invest. *60*, 1266 (1977)
106. Babior, B. M., Kipnes, R. S.: Blood *50*, 517 (1977)
107. Gabig, T. G., Kipnes, R. S., Babior, B. M.: J. Biol. Chem. *253*, 6663 (1978)
108. Tauber, A. I., Goetzl, E. J.: Biochem. *18*, 5576 (1979)
109. Light, D. R., et al.: Biochem. *20*, 1468 (1981)
110. Babior, B. M., Peters, W. A.: J. Biol. Chem. *256*, 2321 (1981)
111. Segal, A. W., Jones, O. T. G.: Nature *276*, 515 (1978)
112. Segal, A. W., Jones, O. T. G.: Biochem. J. *182*, 181 (1979)
113. Segal, A. W., Jones, O. T. G.: Biochem. Biophys. Res. Commun. *92*, 710 (1980)
114. Segal, A. W., Jones, O. T. G.: The cytochrome *b* component of the microbicidal oxidase system of human neutrophils, in: Biological and Clinical Aspects of Superoxide and Superoxide Dismutase, (eds.) Bannister, W. H., Bannister, J. V., pp. 231, New York, Elsevier 1980
115. Segal, A. W., et al.: Lancet *2*, 446 (1978)
116. Hamers, M. H., et al.: Cytochrome *b* and the superoxide-generating system in human neutrophils, in: Biological and Clinical Aspects of Superoxide and Superoxide Dismutase, (eds.) Bannister, W. H., Bannister, J. V., pp. 242, New York, Elsevier 1980
117. Badwey, J. A., Karnovsky, M. L.: J. Biol. Chem. *254*, 11530 (1979)
118. Badwey, J. A., et al.: J. Biol. Chem. *256*, 3479 (1981)

119. Rosen, H., Klebanoff, S. J.: J. Clin. Invest. *58*, 50 (1976)
120. Jandl, R. C., et al.: J. Clin. Invest. *61*, 1176 (1978)
121. Gabig, T. G., Bearman, S. I., Babior, B. M.: Blood *53*, 1193 (1979)
122. Baehner, R. L., Boxer, L. A., Davis, J.: Blood *48*, 309 (1976)
123. Johnston, R. B., Jr., et al.: J. Clin. Invest. *55*, 1357 (1975)
124. Baehner, R. L., et al.: J. Clin. Invest. *56*, 571 (1975)
125. Kakinuma, K., et al.: J. Biochem. (Tokyo) *86*, 87 (1979)
126. Beauchamp, C., Fridovich, I.: J. Biol. Chem. *245*, 4641 (1970)
127. Tauber, A. I., Babior, B. M.: J. Clin. Invest. *60*, 374 (1977)
128. Tauber, A. I., Gabig, T. G., Babior, B. M.: Blood *53*, 666 (1979)
129. Drath, D. B., Karnovsky, M. L., Huber, G. L.: J. Appli. Physiol. *46*, 136 (1979)
130. Weiss, S. J., Rustagi, P. K., LoBuglio, A. F.: J. Exp. Med. *147*, 316 (1978)
131. Klebanoff, S. J., Rosen, H.: J. Exp. Med. *148*, 490 (1978)
132. Takanaka, K., O'Brien, P. J.: FEBS Lett. *110*, 283 (1980)
133. Repine, J. W., et al.: J. Clin. Invest. *64*, 1642 (1979)
134. Tsan, M-F., Chen, J. W.: J. Clin. Invest. *65*, 1041 (1980)
135. Rosen, H., Klebanoff, S. J.: J. Clin. Invest. *64*, 1725 (1979)
136. Green, M. R., et al.: FEBS Lett. *100*, 23 (1979)
137. Finkelstein, E., Rosen, G. M., Rauckman, E. J.: Arch. Biochem. Biophys. *200*, 1 (1980)
138. Allen, R. C., et al.: Biochem. Biophys. Res. Commun. *60*, 909 (1974)
139. Allen, R. C., Stjernholm, R. L., Steele, R. H.: Biochem. Biophys. Res. Commun. *47*, 679 (1972)
140. Chesen, B. D., et al.: J. Clin. Invest. *58*, 789 (1976)
141. Andersen, B. R., Lint, T. F., Brendzel, A. M.: Biochim. Biophys. Acta *542*, 527 (1978)
142. Krinsky, N. I.: Science *186*, 363 (1974)
143. Rosen, H., Klebanoff, S. J.: J. Biol. Chem. *252*, 4803 (1977)
144. Held, A. M., Hurst, J. K.: Biochem. Biophys. Res. Commun. *81*, 878 (1978)
145. Ushijima, Y., Nakano, M.: Biochem. Biophys. Res. Commun. *93*, 1232 (1980)
146. Yost, F. J., Jr., Fridovich, I.: Arch. Biochem. Biophys. *161*, 395 (1974)
147. Sagone, A. L., Jr., King, G. W., Metz, E. N.: J. Clin. Invest. *57*, 1352 (1976)
148. Rosen, H., Klebanoff, S. J.: J. Exp. Med. *149*, 27 (1979)
149. Mandell, G. I.: J. Clin. Invest. *55*, 561 (1975)
150. Hassan, H. M., Fridovich, I.: J. Biol. Chem. *254*, 10846 (1979)
151. Salin, M. L., McCord, J. M.: Free radicals in leukocyte metabolism and inflammation, in: Superoxide and Superoxide Dismutases, (eds.) Michelson, A. M., McCord, J. M., Fridovich, I., pp. 257, London, Academic Press 1977
152. Petrone, W. F., et al.: Proc. Natl. Acad. Sci. USA *77*, 1159 (1980)
153. McCord, J. M., Wong, K.: Phagocyte-produced free radicals: Roles in cytotoxicity and inflammation, in: Oxygen Free Radicals and Tissue Damage, (ed.) Fitzsimons, D. W., pp. 343, Amsterdam, Excerpta Medica 1979
154. Klebanoff, S. J.: J. Biol. Chem. *249*, 3724 (1974)
155. Bass, D. A., Szejda, P.: J. Clin. Invest. *64*, 1415 (1979)
156. Bass, D. A., Szejda, P.: J. Clin. Invest. *64*, 1558 (1979)
157. Diamond, R. D., Clark, R. A., Haudenschild, C. C.: J. Clin. Invest. *66*, 908 (1980)
158. Kazura, J. W., et al.: J. Clin. Invest. *67*, 93 (1981)
159. Salin, M. L., McCord, J. M.: J. Clin. Invest. *56*, 1319 (1975)
160. Roos, D., et al.: J. Clin. Invest. *65*, 1515 (1980)
161. Baehner, R. L., et al.: Blood *50*, 327 (1977)
162. Clark, R. A., Klebanoff, S. J.: Blood *50*, 65 (1977)
163. Baehner, R. L., Nathan, D. G., Castle, W. B.: J. Clin. Invest. *50*, 2466 (1971)
164. Weiss, S. J., LoBuglio, A. F.: Blood *55*, 1020 (1980)
165. Weiss, S. J.: J. Biol. Chem. *255*, 9912 (1980)
166. Weiss, S. J.: Personal communication
167. Katz, P., et al.: J. Clin. Invest. *65*, 55 (1980)
168. Sacks, T., et al.: J. Clin. Invest. *61*, 1161 (1978)
169. Hafeman, D. G., Lucas, Z. J.: J. Immunol. *123*, 55 (1979)
170. Clark, R. A., Klebanoff, S. J.: J. Immunol. *119*, 1413 (1977)

171. Clark, R. A., Klebanoff, S. J.: J. Immunol. *122*, 2605 (1979)
172. Clark, R. A., Klebanoff, S. J.: J. Exp. Med. *141*, 1442 (1975)
173. Slivka, A., LoBuglio, A. F., Weiss, S. J.: Blood *55*, 347 (1980)
174. Nathan, C. F., et al.: J. Exp. Med. *149*, 84 (1979)
175. Nathan, C. F., et al.: J. Exp. Med. *149*, 100 (1979)
176. Levine, P. H., et al.: J. Clin. Invest. *57*, 955 (1976)
177. Clark, R. A., Klebanoff, S. J.: J. Immunol. *124*, 399 (1980)
178. Webb, L. S., Keele, B. B., Jr., Johnston, R. B., Jr.: Infect. Immun. *9*, 1051 (1974)
179. Kakinuma, K., et al.: FEBS Lett. *102*, 38 (1979)
180. McPhail, L. C., DeChatelet, L. R., Johnston, R. B., Jr.: J. Clin. Invest. *63*, 648 (1979)
181. Hatch, G. E., Gardner, D. E., Menzel, D. B.: J. Exp. Med. *147*, 182 (1978)
182. Harvath, L., Amirault, H. J., Andersen, B. R.: J. Clin. Invest. *61*, 1145 (1978)
183. Harvath, L.: Infect. Immun. *25*, 473 (1979)
184. Allred, C. D., Margetts, J., Hill, H. R.: Biochim. Biophys. Acta *631*, 380 (1980)
185. Oyanagui, Y.: Biochem. Pharmacol. *25*, 1465 (1976)
186. Perez, H. D., Weksler, B. B., Goldstein, I. M.: Inflammation *4*, 313 (1980)
187. Carp, H., Janoff, A.: J. Clin. Invest. *63*, 793 (1979)
188. Carp, H., Janoff, A.: J. Clin. Invest. *66*, 987 (1980)
189. Clark, R. A., et al.: J. Biol. Chem. *256*, 3348 (1981)
190. Clark, R. A., Klebanoff, S. J.: J. Clin. Invest. *64*, 913 (1979)
191. Gordon, L. I., et al.: J. Clin. Invest. *64*, 226 (1979)
192. Mandell, G. L., Hook, E. W.: J. Bacteriol. *100*, 531 (1969)
193. Curnutte, J. T., Whitten, D. M., Babior, B. M.: New Engl. J. Med. *290*, 593 (1974)
194. Borregaard, N., Johansen, K. S., Esmann, V.: Biochem. Biophys. Res. Commun. *90*, 214 (1979)
195. Weening, R. S., Wever, R., Roos, D.: J. Lab. Clin. Med. *85*, 245 (1975)
196. Hohn, D. C., Lehrer, R. I.: J. Clin. Invest. *55*, 707 (1975)
197. Mills, E. L., Rholl, K. S., Quie, P. G.: J. Clin. Invest. *66*, 332 (1980)
198. Baehner, R. L., Nathan, D. G., Karnovsky, M. L.: J. Clin. Invest. *49*, 865 (1970)
199. Johnston, R. B., Jr., Baehner, R. L.: Blood *35*, 350 (1970)
200. Misra, H. P., Fridovich, I.: J. Biol. Chem. *247*, 6960 (1972)
201. Winterbourn, C. C., McGrath, B. M., Carrell, R. W.: Biochem. J. *155*, 493 (1976)
202. Brunori, M., et al.: Eur. J. Biochem. *53*, 99 (1975)
203. Demma, L. S., Salhany, J. M.: J. Biol. Chem. *252*, 1226 (1977)
204. Goldberg, B., Stern, A.: J. Biol. Chem. *251*, 6468 (1976)
205. Goldberg, B., Stern, A.: Biochim. Biophys. Acta *437*, 628 (1976)
206. Goldberg, B., Stern, A.: Arch. Biochem. Biophys. *178*, 218 (1977)
207. Sutton, H. C., Roberts, P. B., Winterbourn, C. C.: Biochem. J. *155*, 503 (1976)
208. Lynch, R. E., Lee, G. R., Cartwright, G. E.: J. Biol. Chem. *254*, 1015 (1976)
209. Lynch, R. E., Thomas, J. E., Lee, G. R.: Biochem. *16*, 4563 (1977)
210. Kellogg, E. W., III, Fridovich, I.: J. Biol. Chem. *252*, 6721 (1977)
211. Lynch, R. E., Fridovich, I.: J. Biol. Chem. *253*, 1838 (1978)
212. Bartosz, G., et al.: Eur. J. Biochem. *73*, 261 (1977)
213. Fee, J. A., Teitelbaum, H. D.: Biochem. Biophys. Res. Commun. *49*, 150 (1972)
214. Fee, J. A., Bergamini, R., Briggs, R. G.: Arch. Biochem. Biophys. *169*, 160 (1975)
215. Lynch, R. E., Fridovich, I.: J. Biol. Chem. *253*, 4697 (1978)

The Flavin Redox-System and Its Biological Function

Franz Müller

Department of Biochemistry, Agricultural University, De Dreijen 11,
NL-6703 BC Wageningen, The Netherlands

Table of Contents

Abbreviations

ESR, electron spin resonance
NMR, nuclear magnetic resonance
ENDOR, electron-nuclear double resonance
FAD, flavin adenine dinucleotide
FMN, riboflavin-5'-monophosphate (flavin mononucleotide)
TARF, tetracetylriboflavin
Rfl, riboflavin
Lfl, lumiflavin

1 Introduction

More than one century ago a yellow, fluorescent pigment was isolated from whey by Blyth [1]. In the subsequent years yellow pigments were extracted from various biological materials. Depending either on the source of isolation or the physical appearance, these natural products were named e.g. "lactochrome", "lycochrome", "ovoflavin", "lactoflavin", "hepatoflavin", or "verdoflavin". Later, it became evident that all these compounds are riboflavin (vitamin B_2).

Some highlights of the developments in the field of flavins and flavoproteins are presented in Table 1. This table is an extension of a similar table published by Beinert [56] who summarized the progress in the field achieved up to 1958. Due to the wealth of papers published during the last two decades, the discoveries listed in Table 1 should not be considered as a comprehensive overview but rather as a selection of a few topics reflecting the author's own view. Looking back at the "history" of flavin research, it is striking that the great activities displayed especially during the fourth decade of this century came to an almost complete stop in the subsequent two decades. One reason for the decreasing interest in the natural product under consideration lies in the fact that the chemists regarded flavins as chemically rather stable compounds. Perhaps the main reason for the decreasing activities in the flavin field was the fact that the isolated compounds from different sources were structurally invariable, not provoking the synthetic skill of an organic chemist. In this context, the potentiometric investigations carried out by Michaelis and his co-workers [57,58] were exceptions. In fact, this work provided the basis for a revival of the interest in flavins in the sixth decade of this century. Beinert [18] investigated the half-reduced flavin system by the ESR technique proving that such a system contains flavin radicals. This observation, on the one hand, and the initiated chemical studies by Hemmerich and Fallab [59], on the other hand, were the great impulses leading flavin research into a new and flourishing era. Of course, these studies were accompanied by many puzzling findings by flavoenzymologists who succeeded in the isolation and characterization of a number of flavoproteins. Since 1965, the activities of the chemists and biochemists in the field of flavin research are coordinated in regularly held international meetings. The close interaction between chemists and biochemists facilitated by these meetings made it possible to achieve a great progress during the last 15 years. The growing and continuous interest of

Table 1. Some Highlights of the Developments in the Field of Flavins and Flavoproteins

Year	Discovery	Research Group	Ref.
1879	Riboflavin ("lactochrome") from whey	Blyth	1)
1908	Oxynitrilase from almonds (a controversal enzyme over a long period of time)	Rosenthaler	2)
1926	Xanthine Oxidase	Dixon and Kodama	3)
1932	"Old yellow enzyme", first protein recognized as flavoenzyme	Warburg and Christian	4)
1934	Synthesis of lumiflavin	Kuhn et al.	5)
1935	Synthesis of riboflavin	Karrer et al.	6)
	Identification of FMN as prosthetic group of "old yellow enzyme"	Theorell	7)
1936	Synthesis of FMN	Kuhn et al.	8)
1938	Identification of FAD as prosthetic group of D-amino acid oxidase	Warburg and Christian	9)
1939	Lipoamide dehydrogenase from heart	Straub	10)
1950	FAD exists as an intramolecular complex	Weber	11)
1952	Lipoamide dehydrogenase is a constituent of the multi-enzyme pyruvate dehydrogenase complex	Sanadi et al.	12)
1953	Xanthine oxidase contains FAD, Fe and Mo	Green and Beinert	13)
1954	Synthesis of FAD	Christie et al.	14)
	Lactate oxidase, internal flavoprotein mono-oxygenase	Sutton	15)
1955	Succinate dehydrogenase contains a covalently bound flavin	Kearney and Singer	16)
1956	Cytochrome b_2, contains FMN and proto-heme IX.	Boeri and Tosi	17)
	Proof of the existence of flavin-free radicals by the electron spin resonance technique	Beinert	18)
1958	Butyryl-CoA dehydrogenase: formation of long wavelength absorbing species	Steyn-Parvé and Beinert	19)
1959	NADPH-cytochrome c reductase from liver microsomes	Lang and Nason	20)
1960	Isolation of a FAD-containing peptide after proteolysis of succinate dehydrogenase	Kearney	21)
1962	Salicylate hydroxylase, external flavoprotein monooxygenase	Katagiri et al.	22)
1963	Structure and function of oxynitrilase	Becker et al.	23)
1964	The formation of flavosemiquinone metal chelates	Hemmerich	24)
	Binding of chemically modified flavocoenzymes by apoflavoproteins	McCormick et al.	25)
1966	Flavin-10a-hydroperoxides as an intermediate in the reaction of reduced flavin with O_2.	Mager and Berends	26)
	Flavoproteins form neutral and anionic semiquinones	Massey and Palmer	27)
1967	X-ray structure analysis of flavin	Kierkegaard et al.	28)
	Bacterial NADPH-sulfite reductase contains FMN, FAD and Fe as prosthetic groups	Siegel et al.	29)
	Flavosemiquinone anion	Ehrenberg et al.	30)
1968	Microsomal amine oxidase	Machinist et al.	31)
1969	Structure of covalently bound prosthetic group of succinate dehydrogenase	Hemmerich et al.	32)
	Crystal structure of *Clostridium* flavodoxin	Ludwig et al.	33)

Table 1 (continued)

Year	Discovery	Research Group	Ref.
	Structure of sulfite complex of flavoproteins and the possible reaction of flavin hydroperoxides	Massey et al. Müller and Massey	[34] [35]
1970	X-Ray study on cytochrome b_2	Monteilhet and Risler	[36]
1971	Partial sequence of *clostridial* flavodoxins	Tanaka et al.	[37]
	Physical evidence of the existence of flavin hydroperoxides	Müller	[38]
1972	Three-dimensional structure of *Desulfovibrio vulgaris*	Watenpaugh et al.	[39]
	Proposal of a specific interaction between a prosthetic group and apoflavoprotein to explain the various catalytic properties of flavoenzymes	Müller	[40]
1973	8-Hydroxy- and 6-hydroxy-FAD, new prosthetic groups	Ghisla and Mayhew	[41]
	Roseoflavin	Miura et al.	[42]
	Formation of an adduct at N(5) of the prosthetic group in D-amino acid oxidase	Porter et al.	[43]
1974	Fluorescence properties of reduced flavins and flavoproteins	Ghisla et al.	[44]
1975	Three-dimensional structure of p-hydroxybenzoate hydroxylase from *Pseudomonas fluorescens*	Drenth et al. Wierenga et al.	[45] [46]
	Crystal structure of glutathione reductase	Schulz et al.	[47,48]
	X-Ray of flavodoxin from *Anacystic nidulans*	Smith et al.	[49]
1976	Polymer-bound flavins	Shinkai et al.	[50]
1978	Modified flavins as active-site probes of flavoproteins	Moore et al.	[51]
	5-Deazaflavin derivative as a natural product	Eirich et al.	[52]
1979	Covalently bound non-coenzyme phosphorous residues in flavoproteins	Edmondson and James	[53]
1980	Photoelectron spectra of flavins	Eweg et al.	[54]
	Ab initio calculations for flavins	Palmer et al.	[55]

young researchers in the field of flavins and flavoproteins reflects the attractiveness and excitement a single molecule can provide.

In the following the basic chemical and physical properties of free flavin will be described in some detail, because the knowledge of these properties is the key for a detailed understanding of the function of flavoproteins. In addition, some general and common properties of the classes of flavoproteins will be presented and discussed in relation to some new concepts. It will however not be possible to cover the whole literature. For the reader interested in more details, the recent review papers by Massey and Hemmerich [60], Bruice [61], Walsh [62], Simondson and Tollin [63], and Hemmerich [64] (and references therein) should be consulted. An overview of the pertinent research on flavins and flavoproteins is easily accessible by the proceedings of the international symposia [65-72].

2 Physical and Chemical Properties of Common and Modified Flavocoenzymes

2.1 Common Flavocoenzymes

The most common flavocoenzymes associated with biological materials are Rfl, FMN and FAD (Scheme 1). The side chains ribityl, ribityl phosphate and ribityl adenine diphosphate are important constituents of flavocoenzymes since these side chains are in general responsible for the selective binding to a particular flavoapoprotein. The isoalloxazine ring is the redox active chemical entity of the flavocoenzymes. The term "isoalloxazine" is widely used in the literature although it is an unfortunate creation originating from the use of alloxan in the early flavin synthesis [73]. In this terminalogy, 7,8-dimethylisoalloxazine means flavin. According to international recommendations flavin is defined as 7,8-dimethylbenzo[g]-pteridine-2,4(3H, 10H)-dione. It should also be noted that the numbering given in Scheme 1 is the one internationally now accepted but an older numbering is still sometimes found in the modern literature. The terms FMN (flavin mononucleotide) and FAD are also in discrepancy with the common chemical nomenclature because the side chains are not linked to the flavin moiety via a glycosidic bond which can, therefore, not be cleaved by hydrolysis. These terms are, however, internationally accepted.

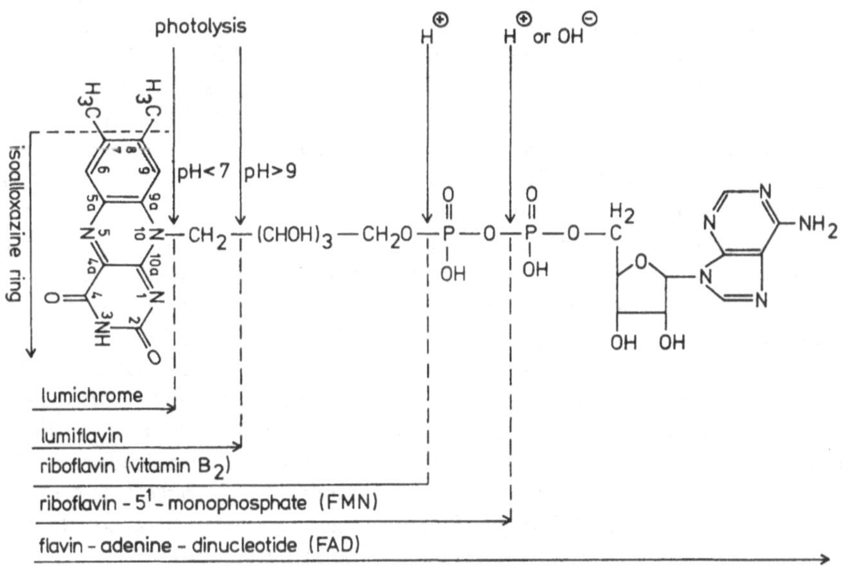

Scheme 1

Hydrolysis of FAD leads to FMN and/or Rfl (Scheme 1). Depending on the pH, free FMN and Rfl can be degraded to lumiflavin or lumichrome by photolysis (Scheme 1). Mainly for this reason lumiflavin serves often as a substitute for the flavocoenzymes in model bio-organic chemistry. Free FAD is much less liable to photolysis, due to the formation of an intramolecular complex [11] which is responsible for the very efficient radiationless deactivation of the flavin in the excited state. Bound to apoflavoproteins all flavocoenzymes are generally quite stable against photolysis although continuous irradiation of flavoprotein solutions over a long period of time may lead to the destruction of the flavocoenzyme and more probably to irreversible damage of the apoflavoprotein.

Riboflavin is synthesized by all green plants and by most microorganisms. A recent account on the pathways of riboflavin biosynthesis is found in Refs. [74,75]. Riboflavin is the precursor of FMN and FAD. Higher animals have lost the capacity to synthesize Rfl and their daily requirements must be supplied by foodstuffs. The biological synthesis of FMN from Rfl is catalyzed by the enzyme flavokinase. The biological transformation of FMN into FAD requires the enzyme FAD synthetase. Both reactions depend on ATP and Mg^{++} as co-factors. More recently, Walsh et al.[76] discovered a flavokinase/FAD synthetase system in *Brevibacterium ammoniagenes* capable of phosphorylating and adenylating many artificial riboflavin derivatives. This finding is of practical use in the small-scale preparation of chemically modified FMN of FAD derivatives. Such derivatives have been used in the past in a few cases as substitutes for the natural prosthetic group of flavoproteins to test the specificity of the coenzyme binding site of a particular flavoprotein.[25,77,78] This technique received much attention in the last few years. Since the discovery of 6- and 8-hydroxyflavins [79,80] (cf. below) the potential of modified flavocoenzymes as active-site probes of flavoproteins became evident. Concomitantly, the interest in the preparation of modified flavocoenzymes grew. In addition, earlier studies [25] revealed that, in general, apoflavoproteins are capable of binding modified flavocoenzymes as long as the demand for a specific side chain is fullfilled. Of course, the replacement of the natural prosthetic group by a modified one affects the catalytic properties of the enzyme studied. The chemical synthesis of riboflavin derivatives has been worked out and such derivatives are easily available in pure form [81-84]. The procedure to phosphorylate riboflavin derivatives on a preparative scale has recently been improved [85]. These preparations, and also commercial FMN, contain a considerable amount of riboflavin phosphate isomers, which are difficult to separate by column chromatography. This problem is emphasized in the chemical synthesis of FAD [14,86,87] where the yield is rather low (20–25%). In this context, it is surprising that a modification of the synthesis of FAD from FMN published by Cramer and Neuhoeffer [88] has not been noticed by workers in the flavin field. According to Cramer and Neuhoeffer, the yield of the chemical synthesis of FAD is drastically improved (~70% pure FAD). The procedure was successfully applied in the author's own laboratory (yield ~60–70%). It is expected that the improved procedure of the FAD synthesis will stimulate the active-site directed studies on flavoproteins because the problem of separating FMN or FAD from their synthetic by-products has already been solved by use of FMN- or FAD-specific affinity column [89,90].

2.2 Modified Flavocoenzymes

The discoveries of Theorell [91] and Warburg and Christian [92] demonstrated that the prosthetic group of flavoenzymes can be released reversibly without proteolytic digestion of the protein. Most flavoproteins known today show a similar behaviour. Although a very strong interaction of binding can occur, the prosthetic group is usually released from the flavoprotein under relatively mild denaturing conditions. We know now that the tight binding is a result of combined effects of electrostatic and hydrophobic interactions between the prosthetic group and the apoflavoprotein. In 1955, Green et al. [93] discovered that only part of the total tissue flavin was released upon treatment with the usual techniques and concluded from this observation that part of the flavin was covalently bound. This proposal was supported by the finding of Kearney [21] who succeeded in isolating a peptide on the level of lumiflavin after proteolytic digestion of succinate dehydrogenase and alkaline photolysis of the isolated flavin peptide. Although this seemed a promising approach to the solution of the problem, ultimately it led to the elucidation of the structure of the covalently bound prosthetic group, the unusual chemical properties of the new flavin prevented a rapid progress in this field. Later, Chi et al. [94] demonstrated that the peptide was not linked to the N(3) position of the flavin. These two results already indicated that the flavin-peptide linkage must occur in the benzene subnucleus of the flavin. In 1966, an identical conclusion was deduced from model studies [84]. Strong evidence of the linkage between the peptide and the 8α-position of the flavin was presented in 1969 [32] and finally proved in 1972 [95].

Today, four kinds of 8α-substituted peptides are known (Scheme 2). The published data up to 1976 were summarized recently [96,97]. The chemical synthesis of the various flavin peptides and their physical properties are described in the references given in Table 2 and were reviewed recently [64].

The amino acid residues of the apoflavoproteins (Scheme 2, (*1*) to (*4*)) involved in the covalent binding to the 8α-position of the prosthetic group possess functional groups which are ionizable in the pH range 6–10 as free molecules. These pK

Table 2. Flavoproteins Containing a Covalently Bound Prosthetic Group[a)]

Enzyme	Covalently Bound Prosthetic Group	Ref.
Cholesterol oxidase	8α-N(1)-Histidyl-FAD	107)
Thiamine dehydrogenase	8α-N(1)-Histidyl-FAD	108)
Cyclopiazonate oxidocyclase	8α-N(1)-Histidyl-FAD	109)
L-Gulonolactone oxidase	8α-N(1)-Histidyl-FAD	110)
L-Galactonolactone oxidase	8α-N(1)-Histidyl-FAD	111)
6-Hydroxy-D-nicotine oxidase	8α-N(3)-Histidyl-FAD	112)
Sarcosine dehydrogenase	8α-N(3)-Histidyl-FAD	113)
Succinate dehydrogenase from beef hart mitochondria	8α-N(3)-Histidyl-FAD	114)
Succinate dehydrogenase from *Vibrio succinogenes*	8α-N(3)-Histidyl-FAD	115)
Choline oxidase	8α-N(3)-Histidyl-FAD	116)
p-Cresol methylhydroxylase	8α-0-Tyrosyl-FAD	117)
Monoamine oxidase	8α-S-Cysteinyl-FAD	118)
Flavocytochrome c_{552} from *Chromatium*	8α-S-Cysteinyl-FAD	119)
Flavocytochrome c_{553} from *Chlorobium thiosulfatophilum*	8α-S-Cysteinyl-FAD	120)
Trimethylamine dehydrogenase	6-S-Cysteinyl-FMN	121,122)

a) For structures, see Scheme 2

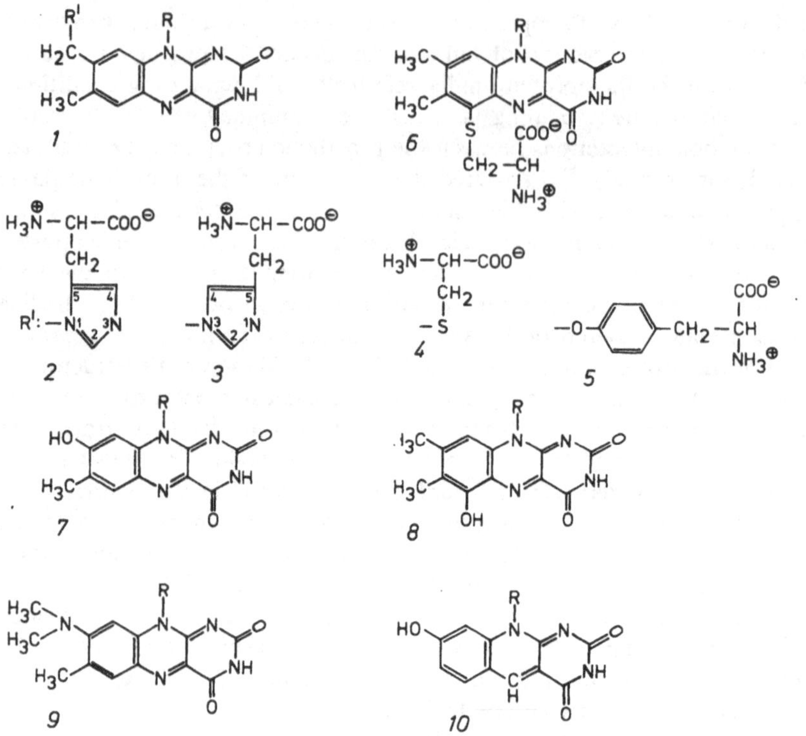

Scheme 2

values may be shifted to even lower pH values in a polypeptide chain. These amino acid residues are susceptible to nucleophilic substitution reactions and could, therefore, give a possible hint at the mechanism of the biological synthesis of flavin peptides. The reaction mechanisms by which flavin is attached to the proteins is still unknown. Some recent works on this subject, however, indicate that naturally occurring flavocoenzymes can be incorporated into the apoflavoproteins. This has been demonstrated by a cell-free synthesis of 6-hydroxy-D-nicotine oxidase [98]. This finding may surprise scientists not familiar with the chemical reactivity of flavin. The reactivity of flavin towards various reagents will be discussed below; here we only mention a few relevant reactions related to the 8α-substituted flavocoenzymes. Bullock and Jardetzky [99] showed that the CH_3 group at C-8 selectively exchanges its protons in D_2O at pH 6.8 in phosphate buffer. The methyl group can also be selectively oxidized by NO_2^- in acetic acid to the corresponding acid [100]. This reaction is comparable to the oxidation of p-nitrotoluene to p-nitrobenzoic acid. Isoalloxazine-6,8-disulfonic acid is converted to the corresponding 8-hydroxy-isoalloxazine by the hydroperoxide anion. This reaction leaves the 6-position unaltered which is chemically less reactive [101]. The chemical synthesis of the 8α-modified flavocoenzymes was achieved by the use of 8α-monobromoflavin derivatives as starting materials which are used as alkylating agents for the preparation of the desired amino acid derivatives [102]. These few examples demonstrate the ease with which 8α-substitution of flavin can occur. Therefore, sophisticated biological means to produce covalently bound prosthetic groups in certain flavoproteins are probably not required. This immediately raises the question why some flavoproteins contain covalently bound prosthetic groups and others not. The answer to this question cannot be given at the present time. It is possible that the covalent attachment of the prosthetic group was, at a certain evolutionary stage, the only means of some flavoproteins to accommodate the flavin needed for catalysis. The polypeptide chain in the isoalloxazine binding pocket probably strongly differs in these enzymes from those known today from crystallographic studies (cf. below). This would mean that covalently bound prosthetic groups should be deeply embedded in the protein interior. It is obvious that much more experiments are needed to shed light on this problem and it is probable that only three-dimensional data can answer our questions. To obtain such data will not be an easy task.

There is, up to now, one exception known to the four kinds of the above mentioned covalent linkages. The prosthetic group of trimethylamine dehydrogenase is linked via the C(6)-atom of the flavin to a cysteinyl residue (Scheme 3, (*5*)). As mentioned above the less reactivity of C(6) of flavin as compared to that of $CH_3(8)$ requires probably some chemical modification of the prosthetic by biological means prior to covalent attachment. The C(6)-S-Cysteinyl flavin was synthesized recently starting with 6-nitro flavin [103] which was subsequently reduced to the amino derivative and transformed to the corresponding bromo derivative via diazotation. Reaction of the bromo derivative with cysteine gave the desired 6-S-Cysteinyl derivative [104].

The flavoproteins known to contain covalently-found prosthetic groups are listed in Table 2. In addition to the flavoproteins also carcosine oxidase from *Corynebacterium* contains a covalently-bound FAD [105]. Its mode of linkage is not yet known. Furthermore it seems to be the first enzyme reported to contain equivalent amounts of covalently-bound and dissociable FAD. Djmehtlyglycine dehydrogenase prossesses also a covalently linked prosthetic group, which structure is probably identical with that of (*2*) [106].

The 8α-modified flavins show some remarkable physical properties which differ from those of common flavins. Since these properties are very helpful in the identification of modified 8α-substituted flavocoenzymes originating from biological materials the physical properties are presented briefly here. Compared to riboflavin (Scheme 2, (*1*), R′ = H) the visible light absorption properties of (*2*) to (*5*) are very similar except that the second absorption maximum is blue-shifted by about 20 nm. In (*2*) and (*3*) this blue shift depends on the state of ionization of the imidazole. Similarly, the quantum yield of the fluorescence emission is strongly pH-dependent. This property can be used as a first indication of the structure of the 8α-amino acid residue of peptides obtained proteolytically. Whereas (*1*) exhibits a pH-independent quantum yield in the range between 3–8, (*2*) to (*5*) are weakly fluorescent at pH 7 (about 10% or less of that of (*1*)). At pH 3.4 the fluorescence quantum yield of (*2*) and (*3*) is the same as for riboflavin. This is not valid for (*4*) and (*5*), the fluorescence of which does not increase with decreasing pH. The increase of fluorescence in (*2*) and (*3*) is due to the protonation reaction of the N-atom of imidazole. In fact, from this observation the pK_a values of 5.2 for (*2*) and 4.5 for (*3*) were determined. Therefore, from the pH-dependence of the fluorescence of an unknown sample a good indication of the structure of the 8α-substituent can be obtained. To distinguish e.g. between (*2*) and (*3*) the measurement of the pH-dependent fluorescence quantum yield is routinely applied. Two other properties of (*2*) and (*3*) can be used to distinguish them. Firstly, (*2*) is irreversibly reduced by BH_4^-, whereas (*3*) does not react. Secondly, (*2*) exhibits a much greater storage stability in solution than (*3*). (*4*) and (*5*), on the other hand, can be distinguished from each other and from (*2*) and (*3*) by performic acid oxidation. Upon this treatment the fluorescence of (*4*) increases to the level of riboflavin and shows a pH-independent quantum yield, whereas the fluorescence of (*2*), (*3*) and (*5*) is not affected by this treatment. Thus, (*5*) is the only derivative that fluorescence is not influenced by chemical treatment of pH. Furthermore, it is noticeable that (*2*), (*3*), (*4*) and (*6*) possess redox potentials ($E_{m,7}$) about 30 mV more positive than that of riboflavin (—190 mV), indicating an increase in electron affinity as compared to that of (*1*). The physical and chemical properties of 8α- and 6-substituted riboflavin derivatives were summarized recently [123].

2.3 Structure and Properties of other Modified Flavocoenzymes

In 1971 the discovery of an orange coloured prosthetic group was reported [124]. Because of its unusual colour as a flavin it was called „orange" flavin. The flavin derivative is the prosthetic group of an enzyme present in *Megasphaera elsdenii*. The enzyme was later identified as an electron-transferring flavoprotein [125]. From the purified enzyme variable amounts of 8-hydroxy-FAD [41,126] could be extracted together with FAD and 6-hydroxy-FAD [127,128]. Initially, 6-hydroxy-FAD was named "green" flavin because of its physical appearance to the eye. These unusual prosthetic groups are non-covalently bound to the apoprotein and can, therefore, be removed from the protein by common, mild denaturing procedures. From the total amount of flavin present in the enzyme only a small fraction consists of the modified prosthetic group. In addition, the modified FAD's cannot be detected in bacterial suspensions that are either treated immediately with acid or incubated under anaerobic instead of aerobic conditions [129]. This indicates that the oxydation (hydroxylation) reaction occurs by some reactive oxygen species generated by the enzyme in a biologically non-relevant reaction. The generation of these purification artifacts by other enzymes than the flavoproteins (or by a combined reaction of two or more enzymes) present in the crude extract cannot be excluded, since the purified enzymes do not yield the artifacts. The oxidation of C(6) of flavin is a surprising fact whereas the $C(8)CH_3$ group is relatively easily oxidizable as shown my model studies (see above). Enzyme molecules containing the modified prosthetic group are enzymatically not active.

The structure of the modified flavocoenzymes was elucidated by chemical synthesis and comparative physical studies [126,130] (Scheme 2, (*7*), (*8*)). The compounds possess some unusual properties some of which are collected in Table 3. The most prominent difference between (*7*) and (*8*) is the fluorescence behaviour: (*7*) is fluorescent, (*8*) does not fluoresce. Moreover, at pH $>$ 10 the fluorescence quantum yield of (*7*) increases by a factor of about 2, in contrast to normal flavin the fluorescence of which is quenched. By this property (*7*) and (*8*) are easily distinguished (Table 3). From the visible absorption properties of analogs of (*7*) and (*8*) it was concluded that both compounds can exist in two tautomeric forms (proton on N(1) or C(8)O, C(6)O), leading to quinoid structures.

Table 3. Some Physical and Chemical Properties of Other Modified Flavocoenzymes[a]

Property	Compound[b]			
	(7)	*(8)*	*(9)*	*(10)*
$E_{m,7}$ (mV)	−332	—	−222	−373
pK_a	4.8 (C(8)OH)	7.1 (C(6)OH)	1.6 (C(8)N(CH_3)$_2$	6.3 (C(8)OH)
	11.5 (N(3)H)	>9 (N(3)H)	10.8 (N(3)H)	12.2 (N(3)H)
Absorption maximum (nm)				
pH 3.0	435, 262	422, 262	— —	395, 267
pH 7.0	472, 300	— —	505, 314	420, 295[c]
pH 13.0	472, 283	550, 434	492, 308	420, 290
Fluorescence emission:				
maximum (nm)	528	[d]	[d]	not determined
intensity[e]	1.34	—	—	
Source	ETF[f] from *M. elsdenii*	ETF from *M. elsdenii*, glycolate oxidase	*Streptomyces davawensis*	*Methanobacterium*
Ref.	[126,127]	[128]	[131]	[52]

a) Riboflavin derivatives
b) For structures see Scheme 2
c) pH 8.85
d) Non-fluorescent
e) Relative to riboflavin (=1)
f) ETF = electron-transferring flavoprotein

A red coloured flavin derivative was isolated from the culture medium of *Streptomyces davawensis* and was named "roseoflavin" because of its colour [42,131]. Its chemical structure was ascertained by synthesis and photolysis of the natural product [132]. It is a riboflavin derivative possessing an N-dimethyl group at position 8 (Scheme 2, (*9*)). Some properties of (*9*) are given in Table 3. "Roseflavin" is non-fluorescent and very sensitive to light in organic solution. Light catalyzes the elimination of one of the methyl groups at position 8. The thus formed 8-methylamino derivative is fluorescent. Although "roseoflavin" binds to various apoflavoproteins it does not restore the corresponding catalytic activity but it possesses weak antibacterial and antiriboflavin activity [133].

Another Japanese research group isolated two riboflavin derivatives from the culture filtrate of *Schizophyllum commune* and called these natural products "Schizoflavin" 1 and 2 [134,135]. The physical and chemical properties of these compounds are almost identical to those of riboflavin. The only structural difference between riboflavin and the "Schizoflavins" lies in the oxidation of the terminal side chain carbon atom, i.e. alcohol *vs.* aldehyde or carboxylic acid.

Cheeseman et al. [136] reported in 1972 the purification and properties of a fluorescent compound from *Methanobacterium*. Owing to the first visible absorption at 420 nm the unknown compound was called Factor-420. It was not until 1978 till it became clear that the compound is a flavin-like molecule [52]. By different physical techniques it could be shown that the molecule is a 5-deaza-FMN derivative (Scheme 2, (*10*)) where at position 8 the methyl group is replaced by a hydroxy group. In addition, the side chain phosphate group is esterified by a lactyl group which, in turn, is linked to a diglutamyl moiety via a peptide bond. Factor-420 functions as an electron carrier between H_2 (—420 mV) and NADH (—320 mV) (cf. Table 3). As a deazaflavin it shows an unexpected reactivity towards molecular oxygen in the 1,5-dihydro state. 1,5-Dihydrodeazaflavins are, in contrast to 1,5-dihydroflavin, rather stable under aerobic conditions. In contrast, Factor-420 in its 1,5-dihydrostate reacts smoothly with oxygen. The chemical and physical properties of (10) have been confirmed recently by synthesis of a deazaflavin analog [137].

The biological function of Factor-420 is to catalyze the electron transport between hydrogen and pyridine nucleotide in *Methanobacteria* (anaerob). These *Archaebacteria* are obligate anaerobes. In this context some proposals have been put forward with respect to the evolution of biological reactions catalyzed by flavoproteins [138].

3 The Flavin Redox System

The most prominent feature of the chemistry of flavin is its redox properties. These properties make flavin especially suitable for its broad involvement in biological reactions. In the following the pH-dependent species formed in one- and two-electron reductions will be dealt with first, including their visible absorption and fluorescence properties. These physical properties form the basis of many kinetical and analytical studies. In Scheme 3 the structures refer to the free and protein-bound prosthetic groups (cf. Scheme 1). To study the physical properties of the flavocoenzymes often N(3)-alkylated lumiflavin (R = CH_3) is used which is better soluble in a variety of solvents. Other physical and chemical properties of these species will be discussed subsequently.

3.1 Flavoquinone

In all three redox states, flavin is an amphoteric molecule (Scheme 3). Flavoquinone, i.e. oxidized flavin, possesses ionization constants of ~ 0 and ~ 10. It should be noted that the first protonation occurs at N(1), which is the most basic nitrogen in $Fl_{ox}H$ and not N(5). In very strong acidic solutions, however, N(5) can also be protonated reversibly giving the dication. The light absorption properties of flavoquinone species are given in Table 4. Deprotonation at N(3) has little influence on the shape of the light absorption spectrum, except that the second band is shifted hypsochromically. The monocation exhibits an apparent single-band visible absorption spectrum which is resolved in cations carrying alkyl substituents at N(1) or C(2α) [140]. A similar behaviour is shown by the dication the maximum of which is shifted bathochromically by 50 nm as compared to that of the monocation. The light absorption spectrum of (*12*) depends strongly on the polarity of the solvent [139–141]. Especially the second absorption band shows an unexpected behaviour. Very recent data [142] indicate that the second transition in the flavin spectrum must be assigned to a mixed state containing anamalous n — π* character arising from the anomalous behaviour of the N(1) eone pair. (*11*) and (*12*) are fluorescent, whereas (*13*) is non-fluorescent because of quenching by OH^-. In acidic solution (*11*) is also non-fluorescent due to quenching by H^+ but quaternary flavinium salts in organic solvents

Table 4. Comparison of the Light Absorption Characteristics of the Amphoteric Species of Lumiflavin in the Three-Redox States

Species [a)]	Solvent	Absorption maxima (nm) (Molar extinction coefficient, mM^{-1} cm^{-1})	Ref.
$Fl_{ox}H$	pH 7	445(12.4), 367(10.0), 266(42.2)	139–141)
Fl_{ox}^-	pH 12	447(12.0), 352(11.6)	139)
$Fl_{ox}H_2^+$	6N HCl	390(21.0)	139, 140)
$Fl_{ox}H_3^{2+}$	conc. H_2SO_4	440(21.0)	139)
$\dot{F}lH_2$ [b)]	pH 5	580(3.6), 502(3.9), 357(7.7), 326(5.8)	150)
$\dot{F}lH_2$ [b)]	$CHCl_3$	642(4.4), 603(4.0), 490(2.4), 460(2.0), 385(5.9), 326(7.2)	150)
$\dot{F}lH^-$	DMF	477(5.9), 400(11.5), 373(17.6)	30)
$(\dot{F}lHZn)^+$ [c)]	DMF	810(0.035), 537(3.6), 502(3.9), 395(12.3), 375(15.2)	151)
$\dot{F}lR_2$ [d)]	$CHCl_3$	447(9.0), 368(21.0)	150)
$\dot{F}lH_3^+$	6N HCl	488(7.2), 358(8.0)	139)
$\dot{F}l_{red}H_3$	pH 5	395(2.7), 295(6.8)	139, 153, 154)
$\dot{F}l_{red}H_2^-$	pH 8.5	342(4.0), 285(7.5)	139, 153, 154)
$\dot{F}l_{red}H_4^+$	6N HCl	316(9.5)	139)

a) For structures see Scheme 3
b) N(5)-ethyl derivative
c) Flavosemiquinone chelate
d) Neutral, red flavosemiquinone

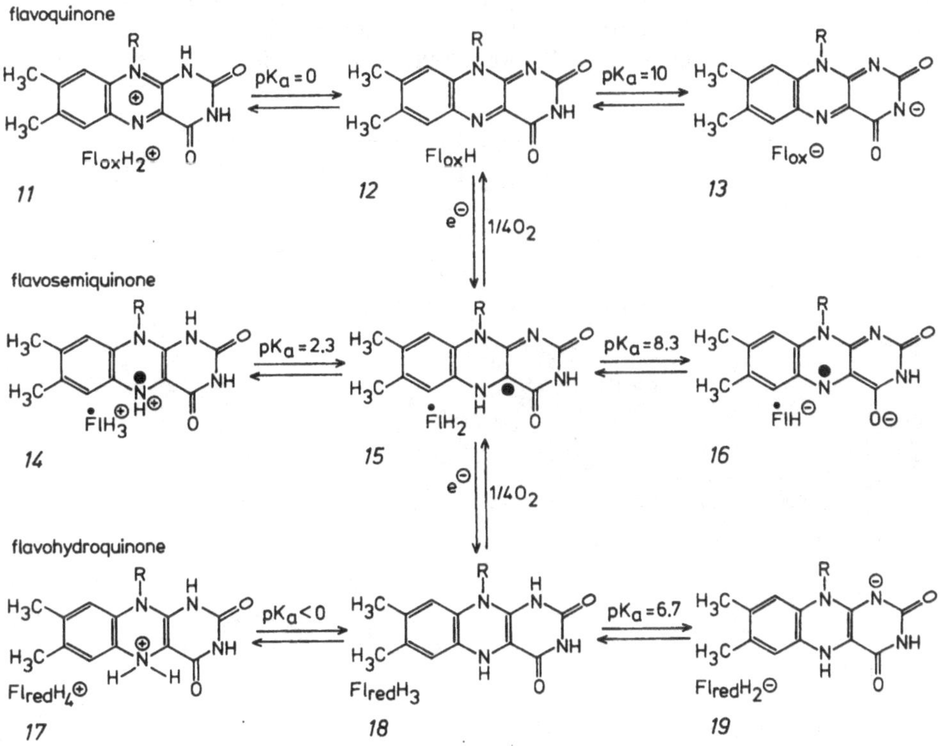

Scheme 3

are fluorescent [140,143]. In contrast to (*12*), Rfl and FMN, FAD is only weakly fluorescent in neutral aqueous solution due to dynamic quenching by the formation of an intramolecular complex [11,144]. Based on this property a procedure was developed to analyze the purity of FAD [145].

3.2 Flavosemiquinone

The addition of one half equivalent of e.g. dithionite to a solution of (*12*) yields a so-called "half-reduced" system. "Half-reduced" is not a very meaningful term, it only implies that one electron is added to flavin without any indication of the spin distribution. This is best demonstrated by equilibrium (1)

$$Fl_{ox}H + Fl_{red}H_3 \rightleftarrows 2\,\dot{F}lH_2 \qquad (1)$$

showing that under unfavourable conditions only a very small concentration of flavosemiquinone may be formed. In neutral aqueous solution the system is strongly disproportionated yielding only a few percent of flavosemiquinone. For this reason Michaelis et al. [57,58] could only infer the existence of the flavosemiquinone. Beinert [56] proved its existence by ESR. The pK value of the deprotonation reaction of (*15*) was determined by ESR [30], potentiometric [146] and pulse radiolysis [147] technique. The pK value of the protonation reaction of (15) was obtained by pulse radiolysis studies [147]. The complexity of a "half-reduced" flavin solution is best demonstrated by the pH-dependent spin concentration of the system as determined by ESR [30] and calculation (Scheme 4). From Scheme 4 it is evident that the flavosemi-

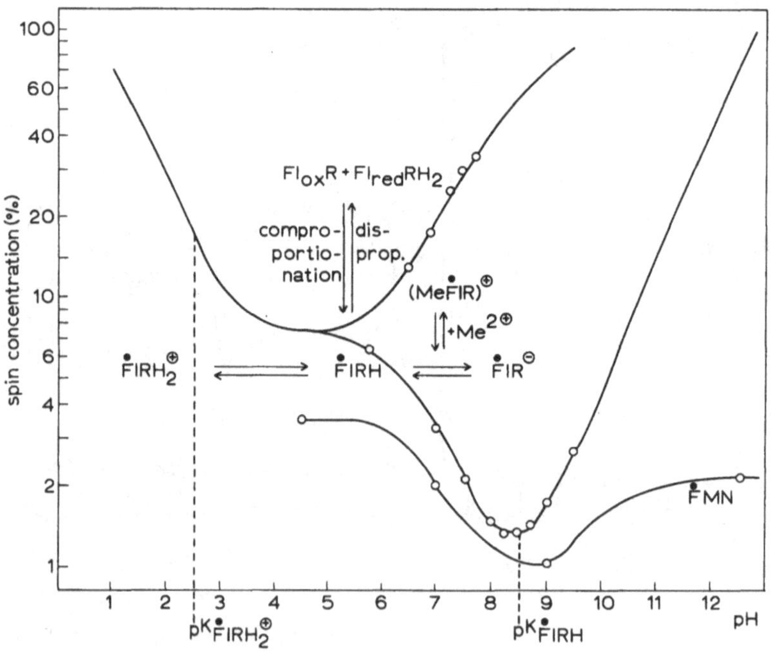

Scheme 4

quinone formation strongly depends on the pH-dependent disproportionation and comproportionation reactions. Theoretically, at low and at high pH values 100% flavosemiquinone formation is expected. This is found for $\dot{F}lH_3^+$ (*14*) but not for $\dot{F}lH^-$ (*16*). The reason for the latter observation is that N(3)-alkylated flavins undergoes hydrolysis at pH > 10. For unsubstituted flavins, e.g. FMN, it is known that monomeric flavin species tend to form sandwich complexes, depending on solvent polarity and ionic strength [148]. On the other hand, the flavosemiquinone cation (*14*) is thermodynamically the most stable one among the flavosemiquinone species. At pH < 0 the "half-reduced" flavin system is fully comproportionated. The high stability of (*14*) made it possible to crystallize it, as done as early as 1937 by Kuhn and Ströbele [149]. The pH-dependence of equilibrium (1), assuming that the effect of formation of bimolecular complexes can be neglected, is given by Eq. (2) [30]

$$k(H) = K_F \frac{(1 + \dot{K}/H)^2}{1 + (K_{ox} + K_{red})/H + K_{ox}K_{red}/H^2} \tag{2}$$

where k(H) is the pH-dependent equilibrium constant belonging to (1), K_F is the semiquinone formation constant among the neutral species ($Fl_{ox}H$, $\dot{F}lH_2$ and $Fl_{red}H_3$) and H is the hydron ion activity. K_{ox}, $\dot{K}$ and K_{red} are the acidity constants of $Fl_{ox}H$, $\dot{F}lH_2$ and $Fl_{red}H_3$, respectively. The theoretical curves in Scheme 4 are calculated using $pK_F = 1.64$, $pK_{red} = 6.3$ and $p\dot{K} = 8.4$ for the N(3)-alkylated system (e.g. $\dot{F}lR^-$) and $pK_F = 2.5$, $pK_{ox} = 10$, $p\dot{K} = 8.6$ and $pK_{red} = 7.0$ for FMN. It should be kept in mind that Scheme 4 applies only to model systems in aqueous solution and a completely different situation is faced with protein-bound flavocoenzymes. Nevertheless, the complexity of the system makes it a very difficult task to decide in model reactions whether the flavosemiquinone or the flavohydroquinone is the primary reaction product using flavoquinone as starting material. The same holds of course studying the mechanism of oxidation of flavohydroquinone. Extreme caution must, therefore, be exercised in the extrapolation of model studies to biological reactions catalyzed by flavoproteins. The biological relevant flavosemiquinones, i.e. (*15*) and (*16*), have been and still are called blue and red radicals. These terms should be avoided because the corresponding flavosemiquinone species have been definitely identified and characterized. The light absorption properties of (*16*) were concealed for a long time although visible spectra of red protein-bound flavosemiquinones were reported [26]. The reason for this is the high disproportionation of partially reduced free flavin in aqueous solution as outlined above. This jeopardized the identification of the visible absorption properties of (*16*). The discovery [30] that an almost quantitative yield of (*16*) could be obtained using an organic solvent (DMF) instead of aqueous solutions made it possible to characterize the flavosemiquinone anion. At the same time [150] it was also found that N(3), C(2α)- or C(2α), C(4α)-dialkylated flavin derivatives yield red coloured, neutral flavosemiquinones. The first absorption maximum of these neutral, red flavosemiquinones is shifted to the blue (~30 nm) as compared to that of (*16*) (Table 4). It is possible that the neutral, red flavosemiquinone can also occur in flavoproteins although the spectra of all red flavoprotein semiquinones known resemble that of (*16*). The colour of solutions of (*15*) vary between green and blue (Table 4), depending on the solvent used. The strong

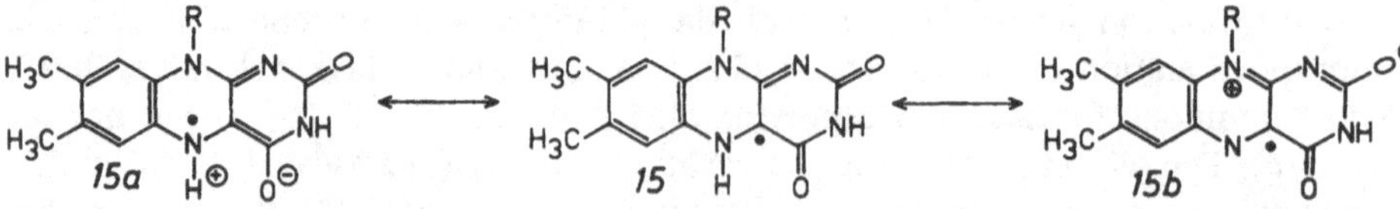

Scheme 5

negative solvatochromicity is explained by the mesomeric structures (*15*) and (*15a*) (Scheme 5). More recently, it was proposed that the mesomeric structure (*15b*) (Scheme 5) contributes also to the stabilization of protein-bound flavosemiquinone [138]. From Scheme 4 it follows that in neutral aqueous solution only a few percent of (*15*) is formed. The spectral properties of (*15*) could, therefore, not be quantified until it was discovered that N(5)-alkylation prevents disproportion of a half-reduced flavin system [152]. The flavosemiquinone cation (*14*) is biologically not relevant but is an ideal species for elucidating the structure of new flavin derivatives of chemical or biological origin because of its quantitative formation and stability towards oxygen. These properties were very useful in studies on the structure of 8α-substituted flavinyl peptides [32].

3.3 Flavohydroquinone

Reduction of flavin by two electrons yields the 1,5-dihydroflavin (Scheme 3), often called "reduced flavin". Since isomeric two-electron reduced flavin structures are known (cf. below), the term "reduced flavin" should be avoided unless defined to prevent misunderstanding. From all flavin species possible in a redox reaction the solution of 1,5-dihydroflavin is, in contrast to that of some isomeric compounds, devoid of a stront colour but not colourless as indicated by the term "leucoflavin", which is still used (Table 4). The only true colourless species is (*17*). Because of the very high oxygen-sensitivity of 1,5-dihydroflavin its chemical and physical properties were investigated only recently [153]. Long before crystallographic data on flavins were available, conclusions were drawn from the molar extinction coefficient at 450 nm of 1,5-dihydroflavins with respect to the planarity of the molecules. From the data presented in Table 5 it was proposed that anionic 1,5-dihydroflavins are less planar than the neutral ones [39]. This suggestion was later fully supported by X-ray studies [155]. The extinction coefficients are indeed very useful values to estimate the planarity of free and protein-bound 1,5-dihydroflavins as recently shown by Ghisla et al. [154]. These authors also found that (*18*) and (*19*) possess weak fluorescence properties at low temperatures [154]. The fluorescence emission maximum appears at 495 nm for (*18*) and at 510 nm for (*19*). At room temperature free 1,5-dihydroflavins are non-fluorescent. In contrast, many 1,5-dihydroflavins bound to apoflavoproteins exhibit weak fluorescence properties at room temperature [154,156].

With respect to (*15*) the possible mesomeric structures are given in Scheme 5. A few remarks will be placed here regarding possible mesomeric or tautomeric structures for the other species depicted in Scheme 3. For all one- or two-electron reduced species various tautomeric forms can exist, e.g. iminol tautomers. Such tautomers do not exist in flavoquinone. In addition, in (*16*) and (*19*) the negative charge could also be placed on the other hetero atoms in the pyrimidine subnucleus

Table 5. Comparison of the Molar Extinction Coefficients at 450 nm of 1,5-Dihydroflavins in the Neutral and Anionic State

Compound	Neutral (pH 2.2)	Anionic (pH 8.5)	Ref.
	ε_{450} mM^{-1} cm^{-1}	ε_{450} mM^{-1} cm^{-1}	
$FMNH_2$	1.39	0.90	[139]
$FADH_2$	1.57	1.06	[139]
$TARFH_2$	0.89	0.62	[139]
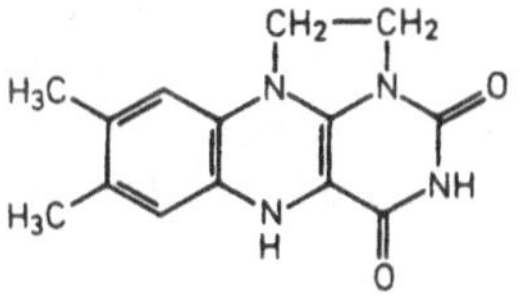	1.58	—	[139]
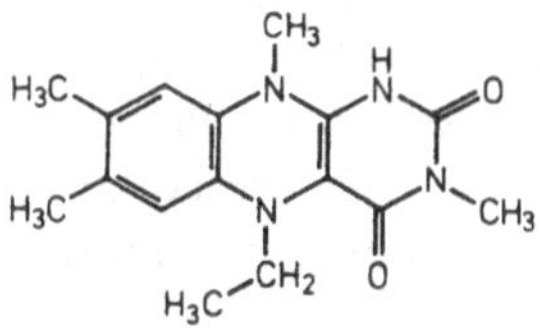	0	—	[153]

of the flavin system, except on N(3). Finally, it is important to note that the formation of all one- or two-electron reduced species (Scheme 3) is thermodynamically fully reversible.

3.4 Flavin Redox Species Bound to Apoflavoproteins

In biological systems all species shown in Scheme 3 can be formed, except the cationic species. The pK value for the formation of *(13)* has been determined for a few flavoproteins, but its formation may lead to the release of the prosthetic group, because either N(3)H is important for the binding interaction with the apoflavoprotein or the alkaline solution (pH ~10) leads to denaturation of the protein. For instance, pK_a values of 8.0 and 9.4 were determined for glycolate oxidase [157] and D-amino acid oxidase [158], respectively. These findings indicate that the N(3)H group of the protein-bound flavoquinone in these two enzymes is in a highly polar environment and probably not hydrogen-bonded to the apoenzymes. This suggestion is supported by the observation that the addition of an inhibitor to these enzymes causes an increase of the pK values by about 2 pH units. This indicates that the environment of the N(3)H group (and probably other parts of the prosthetic group) changes from hydrophilic to more hydrophobic upon binding of inhibitor or substrate (expulsion of water from the binding site) as indicated by the accompanied spectral changes [157, 158]. In this situation the N(3)H may be hydrogen-bonded to the apoenzyme (cf. below).

Flavoproteins form either *(15)* or *(16)* in high yield, if they are formed at all. A helpful, general rule is that flavoprotein oxidases (see below) form *(16)*, flavoprotein dehydrogenases and flavodoxins form *(15)* whereas flavoprotein monooxygenases give no flavosemiquinone. As with all rules, there are exceptions. For instance glucose oxidase is able to form both (*15*) and (*16*) with an ionization constant of 7.3 [27,159]. Although anionic protein-bound flavosemiquinones are not transformed to the neutral one, even at acidic pH (~5), in a few cases it was observed that the addition of an organic, complexing molecule (inhibitor) to e.g. the anionic flavosemiquinone of D-amino acid oxidase yields the neutral flavosemiquinone in a high yield [150]. It should be noticed that the flavoprotein semiquinones observed, which are easily obtained chemically in a photocatalyzed reaction [27,160], may not be of biological relevance. Glucose oxidase, and in fact most flavoprotein oxidases, easily form the semiquinone in a chemical reaction, but nor physical nor chemical evidence could be obtained up to now that the flavosemiquinone is involved in the catalysis [159]. For the dehydrogenases it is known that flavosemiquinone plays an important role in catalysis.

Flavohydroquinone bound to apoproteins plays a very important role in flavoprotein-catalysis, either in the electron-transfer to substrates or other enzymes or in the oxygen activation reaction. The chemical reactivity of 1,5-dihydroflavin bound to apoproteins can differ drastically from that of free flavin. The reactivity is likely governed by factors such as the conformation of the bound flavohydroquinone and the ionization state (cf. below).

4 Chemical Properties of Free and Protein-Bound Flavin

4.1 Flavoquinone: Nucleophilic Reactions

Flavoquinone, the resting state of isolated flavoproteins, was for a long time considered as chemically rather unreactive. The early pioneers in the field studied the alkaline hydrolysis of lumiflavin to prove the chemical structure. Products of hydrolysis were identified as urea and a substituted quinoxaline carboxylic acid [161-163]. This reaction was reinvestigated by Bruice and co-workers [164] under milder conditions. The initial product leads to the formation of an ureido carboxylate by hydrolytic scission between positions 3 and 4 of flavin. For the related alloxazines it was claimed that hydrolytic scission occurs between position 2 and 3 [165]. This was proven to be incorrect [166] so that it can now be concluded that both kinds of molecules are hydrolyzed by the same initial mechanism. The aim of these studies was to determine which positions of flavin are susceptible to nucleophilic reactions because nucleophilic dark reactions may play a role in flavoprotein catalysis. Except for photo-catalyzed reactions and some other nucleophilic dark reactions occurring at the center N(5)-C(4a), which will be discussed below, flavin is only susceptible to nucleophilic reactions after activation of the desired position(s). For instance, nucleophilic reactions occur smoothly at positions 2 and 4 when the oxygen atoms are replaced by sulfur, a methylthio or methoxy group [84,167]. From this it can be concluded that flavoquinone possesses only two functional groups, i.e. N(3)H and C(8)CH_3. Only C(8)CH_3 is chemically an interesting functional group as already demonstrated above. Besides the already mentioned reactions [88-101] C(8)CH_3 condenses with activated aldehydes in a strong acidic medium to deeply coloured products [168]. The dimerization of some flavins to form "bifavins" occurs also via the

$C(8)CH_3$ group [30, 169]. With a halogen at position 8 a variety of 8-substituted flavins can be prepared [170]. This chemical reaction has been applied recently in two destinct ways as a tool to explore the active site of flavoproteins [171]. The high chemical reactivity of 8-chloro-substituted flavocoenzymes was used to fix the prosthetic group in the active center of lipoamide dehydrogenase covalently via the SH group of a cysteine residue [51]. On the other hand, if no such chemical reaction occurs the bound prosthetic group can be tested for the accessibility of the 8-position by reaction with thiolate ions which yields a different flavin chromophore [60]. On the other hand, the chemical transformation of 8-chloroflavin into the corresponding 8-mercaptoflavin by thiolate ions yields a flavin system with pronounced mesomeric and tautomeric structures [171]. Depending on the structure, the first visible absorption maximum occurs between 470 and about 600 nm [171]. In flavoproteins one of the mesomeric or tautomeric structures is generally bound preferentially. From such studies conclusions were drawn with respect to the chemical and physical properties of the flavin domain in several flavoproteins [172]. Although such studies yield some interesting results their interpretation may not be as simple as reported [172].

The finding [43, 173, 174] that flavoproteins form covalent adducts with various compounds stimulated research on the mechanism of flavin-catalyzed oxidations (for recent reviews, see refs. 61, 62). The aim of these model studies is to shed light on the mechanism of biological reactions catalyzed by flavoproteins. The problem has been approached photochemically and by dark reactions, i.e. ground-state chemistry. From the photochemical reactions of flavin with various substrates, yielding generally C(4a)- or N(5)-substituted flavins, it has been concluded that these reactions are models for the covalent addition of carbanions to flavin [175–177] as observed in flavoproteins [174]. The fact that the product of a photochemical reaction of free flavin possesses the same structure (e.g. N(5)-substituted) as that obtained by a dark reaction of a flavoprotein does not allow to conclude that both *mechanisms* are similar, i.e. occur *via* carbanion addition. The photochemical reaction is initiated by the excited flavin in the triplet state. Such reactions are inherently radical in nature [178]. If in such reactions a covalent addition of the carbanion occurs, the available mechanistic studies do not prove it or indicate at what stage of the reaction sequence it would occur. On the other hand, the chemistry of model flavins in the ground state is also very complex [61]. A disadvantage of free flavin as compared to protein-bound flavin is the fact that free flavin in model reactions can undergo complex, secondary reactions (cf. Scheme 3). Principally, it should be possible to mimic the chemistry of flavoproteins in some degree, it is still very difficult to deduce from model studies the biological *relevance* of an observed mechanism. This is simply so because we cannot yet mimic the environment of a protein-bound prosthetic group in its various facets, i.e. microenvironment, specific interactions between flavin and apoprotein, amino acid residues aiding acid-base catalysis. Although the carbanionic mechanism is generally accepted to take place in certain flavoproteins a recent reinterpretation of the reaction mechanism of the oxidation of 1,2-disubstituted substrates by flavins and flavoproteins opens new perspectives [138]. Nevertheless, model studies have revealed that the covalent addition of a substrate to C(4a) of flavin is generally acid-catalyzed [61, 179]. On the other hand, the observation of a radical mechanism of the oxidation of initially formed carbanions by flavoquinone [180] does not imply a similar reaction mechanism in flavoproteins where no evidence could be obtained for the formation of radicals during the catalytic action, e.g. glucose oxidase [159]. In order to circumvent the ambiguities of model studies many research groups have more recently chosen for a different approach to solve the pending problems, i.e. active site directed chemical studies on flavoproteins where compounds related to the structure of natural stubstrates are used [62]. In such reactions the protein-bound flavocoenzyme often forms covalent adducts of different chemical structure [181–184]. These suicide inhibition studies are valuable tools to explore the *chemistry* of the protein-bound flavocoenzyme but it is doubtful if they can contribute much to the unravelling of the catalytic mechanisms of flavoproteins, because the suicide inhibition reactions occur usually only after many catalytic cycles of the enzyme. The suicide reactions have, however, great possibilities for pharmacological applications.

The reaction of sulfite with free flavins, although biologically not relevant, is a useful tool for the characterization of free and protein-bound flavin. The sulfite addition yields a N(5)-substituted flavin with an affinity paralleling the redox potential of the particular flavin analog [35]. This reaction can also be observed in a number of flavoproteins forming the flavosemiquinone anion [34]. This observation led together with other data to a new classification of flavoproteins [34]. Thus, flavoprotein oxidases readily form the sulfite adduct and yield the flavosemiquinone anion upon chemical reduction whereas dehydrogenases and monooxygenases do not react (cf. below). A similar adduct is formed between free flavin and phosphine derivatives [40].

Still another and promising approach to mimic reactions catalyzed by protein-bound flavoquinone is the use of micellar and macromolecular systems [50, 185–187]. In such systems the micro-environment of the flavin can be varied while negatively or positively charged groups can be located in the vicinity of the flavin. Available results indicate that the reactivity of flavin towards various reagents is enhanced in such systems as compared to the free system.

Many flavoproteins contain, besides the flavocoenzyme, also metal ions as prosthetic groups [71]. The study of the interaction between free flavoquinone and d-metal ions is, therefore, of biological interest. In aprotic polar solvents weak complexes are formed [188–190]. Bidentate chelates are formed involving N(5) and the carbonyl group at position 4 of flavin. Complexation causes a red shift of the visible absorption maxima ($\sim$40 nm). From NMR studies [189–190] it has been concluded that delocalization of metal d-electrons towards the ligand is very small and that the spin-transfer mechanism appears to result from indirect σ-π spin polarization at the coordination site. These chelates are probably biologically irrelevant since they would not support flavin-metal electron-transfer.

With respect to flavoquinone it should finally be mentioned that molecular complexes between flavin and aromatic molecules are formed which have their counterparts in flavoproteins. These complexes show often long wavelength absorption, especially when reduced nicotinamides are involved. These deeply coloured species are accepted as charge-transfer complexes, although very convincing evidence has not yet been presented. The most pertinent data were reviewed by Massey and Ghisla [191]. The interaction between a flavin covalently bound to a nicotineamide was studied by Blankenhorn [192]. From a thorough study it was concluded that a hydride-transfer mechanism is involved and that the two reacting molecules are aligned parallel in a sandwich complex, in contrast to a proposal favouring intermediate covalent linkage [193]. X-ray studies on various complexes were conducted [194-198]. The crystallographic data indicate a preferential overlap between the region N(5), C(4a), C(10a) of flavin and the complexing molecule. Such complexes probably play an important role in some flavoprotein-catalyzed reactions.

4.2 Flavosemiquinone: Interactions with Metal Ions

The only flavin species possessing a relatively high metal affinity is flavosemiquinone [24]. In aprotic polar solvents the flavosemiquinone-metal complexes are formed quantitatively and exhibit characteristic visible absorption spectra (Table 4) [151]. In the near infrared region an additional weak absorption is observed. Some solvents, e.g. acetonitrile and pyridine, influence the molar extinction of this absorption. It was therefore tentatively proposed that the near infrared absorption reflects the symmetry of the flavosemiquinone-H_2O-solvent complex. It could be shown that the electronic spin of the radical ligand is also delocalized to the metal ion using highly enriched isotopes of Zn and Cd [199]. From these results it was estimated that the spin density or spin polarization transferred to the nucleus of Cd is about that in the case of Zn. Theses results indicate that semiquinone chelates probably play a role in metal-containing flavoproteins. The stability constant of flavosemiquinone-metal chelates in organic aqueous solutions is about $10^4\ M^{-1}$. Thus, in the presence of metal ions the equilibrium (3)

$$Fl_{ox}H + Fl_{red}H^- \rightleftarrows 2\dot{F}lH^- \xrightleftharpoons{+2Me^{2+}} 2(Me\dot{F}lH)^+ \quad (3)$$

is shifted to the right, i.e. comproportionation occurs (cf. Scheme 4). The spin density distribution in all possible monomeric flavosemiquinone species were determined from the coupling constants of chemically and isotopically substituted flavosemiquinones (Table 6). These results were confirmed by ENDOR measurements [201]. This technique was also successfully applied to protein bound flavosemiquinones where most of the expected coupling constants were observed [212]. From Table 6 it

Table 6. Comparison of the Isotropic Hyperfine Coupling Constants for the Flavosemiquinones

Ring Position	Hyperfine coupling constants in Gauss:			
	(16) Ref. [30]	*(15)* Ref. [152]	*(14)* Ref. [200]	$(Me\dot{F}lR)^+$ Ref. [199]
N-5	7.3	8.0	7.32	7.7
H-5	—	7.6	7.79	—
N-10	3.2	3.6	4.75	3.1
H-10	3.0	3.9	4.97	3.1
H-8	4.0	2.4	3.22	3.9
H-6	3.5	1.7	1.34	3.5
H-9	0.9	—	0.40	—
H-7	—	—	0.50	—

follows that irrespective of the flavosemiquinone species the spin density resides mainly in the central ring of flavin, the highest spin density is always localized at N(5) and C(4a) [203].

The dismutation reaction of a mixture of flavoquinone and flavohydroquinone in DMF was studied kinetically in the presence and absence of redox-inactive metal ions [204, 205]. This system shows a very complex kinetic behaviour with the formation of dimers. The chemical properties of this system resemble somewhat those of FMN in aqueous solution [148]. A partially reduced aqueous flavin solution was also studied by rapid kinetic techniques allowing the determination of the various equilibrium and the corresponding rate constants [206].

The ability of flavosemiquinone to form complexes with amino acids was studied in view of the possible relevance of such complexes in flavoproteins [207]. The results indicate that tryptophan, cysteine and possibly tyrosine could function at the active centers of flavoproteins as agents to stabilize the protein-bound flavosemiquinone [207].

4.3 Flavohydroquinone: Oxygen Activation

Flavohydroquinone possesses no metal affinity but exhibits other interesting chemical properties which were explored by substitution reactions. Flavohydroquinone can, in principle, be substituted at N(1), C(2α), N(3), C(4α), C(4a) and N(5), depending on several factors such as polarity of solvent and nature of the alkylating agent [153]. Protection of N(5) by an acyl group leaves N(1), C(2α), N(3) and C(4α) as alkylating sites [208]. However, N(1) in flavohydroquinone is never alkylated owing to steric hindrance by the peri-substituent at N(10). If N(5) is unprotected it shows a high, unexpected reactivity, being the first position to be alkylated [153]. This reactivity is surprising in view of the fact that the charge in flavohydroquinone anion resides entirely in the pyrimidine subnucleus. In case that N(5) is unprotected C(4a) is also alkylated. Thus, depending on the conditions, either N(5) or C(4a) reacts [153]. Alkylation at C(4a) renders flavohydroquinone redox inactive. The alkyl group can only be removed by a light-catalyzed reaction. The structure of the known dihydroflavins are shown in Scheme 6. Besides the 1,5- (*22*) and 4a,5-dihydroflavin (*21*), three other isomers are known. 1,8-FlH_2

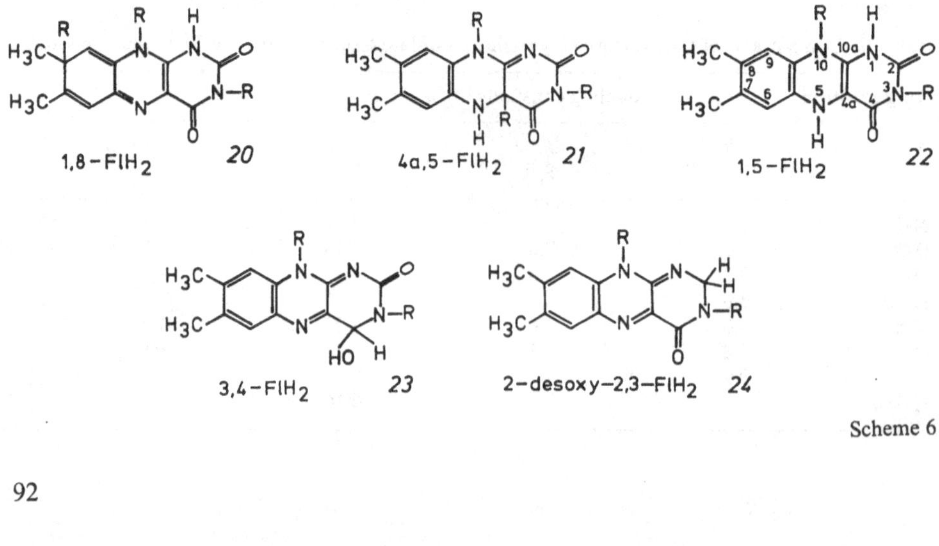

Scheme 6

(20) [209] and 4a,5-FlH_2 *(21)* are only known as alkyl derivatives. Compound *(20)* is only obtained by photochemical reactions. 3,4-dihydroflavin *(23)* is formed by BH_4^--reduction and requires light catalysis [210]. Interestingly, BH_4^--reduction of L- and D-amino acid oxidase yields an identical product formed smoothly in the absence of light [211]. The catalytic properties of these modified enzymes are not influenced indicating that the C(4α) position plays no crucial role in the interaction with the apoprotein and that the redox property of the flavin is not affected. Dihydroflavin *(24)* is formed by BH_4^--reduction of the C(2α), N(3)-dialkylated flavinium salt in a dark reaction. The methoxy group at C(2α) is eliminated by hydride in a secondary reaction [212]. Except *(22)* the dihydroflavins shown in Scheme 6 are stable to air and transformed into flavoquinone by a light-catalyzed reaction. 1,5-Dihydroflavin exhibits rather diffuse visible absorption characteristics (cf. Table 4). In contrast, the derivatives *(20)*, *(21)*, *(23)* and *(24)* show well-defined spectra. The absorption maxima occur at ~430 nm ((*20*)), ~360 nm ((*21*)), ~400 nm ((*23*)), and ~450 nm ((*24*)). Not considering *(22)*, the isomeric 1,5-dihydroflavins are planar, except for the tetrahedral carbon center where the hydride ion is covalently bound.

The most fascinating, but still badly understood, chemical property of flavohydroquinone is the oxygen activation reaction. Mager and Berends concluded from kinetic studies that flavohydroquinone reacts with oxygen forming flavin hydroperoxide as an intermediate [26,213,214]. They postulated the structure of a 10a-flavin hydroperoxide for the intermediate. On the other hand, Hemmerich favoured C(4a) of flavin as the atom of interaction with oxygen [215]. This controversy led subsequently to the postulation of other possible structures, now out of dispute [216]. The observation that certain flavoproteins form sulfite-addition complexes lead Massey and coworkers, considering also other chemical properties of flavoproteins, to interpret the possible oxygen intermediates in terms of their possible biological functions [34]. It was argued, if such intermediates were indeed formed with free and protein-bound flavohydroquinone, that it would be expected that the chemical reaction path from the intermediate to flavoquinone should differ, depending on the function of a particular flavoprotein. In Scheme 7, which is a modification of that originally presented [34], the postulated reaction sequences are shown. Thus, reaction of *(25)* with oxygen can either yield *(26)* or *(27)*, in accord with previous suggestions [213,215]. For convenience, in the following only the possible reaction path of *(26)* is discussed, although the same paths are feasible for *(27)* or even more likely to occur. It was then suggested that flavoprotein oxidases (in the catalytic event oxygen is the terminal electron acceptor) produce hydrogen peroxide. Reduced glucose oxidase, for instance, produces indeed quantitatively hydrogen peroxide upon oxidation, with no indication of intermediate flavosemiquinone formation. Flavoprotein dehydrogenases and flavodoxins are in general making use of the flavosemiquinone in the catalytic reaction. This fact was taken as support for the postulation of the reaction sequence *(26)*→*(28)*→*(29)*. This sequence was observed experimentally with free flavins, flavodoxin and other flavoproteins where the formation of superoxide was proven [217]. Finally, the sequence *(26)*→*(30)*→*(29)* was suggested for the reaction mechanisms of flavoprotein hydroxylases (monooxygenases). Also these enzymes form no flavosemiquinone, either chemically or catalytically. This proposed mechanism is very likely to occur with e.g. p-hydroxybenzoate hydroxylase since ortho-substitution of

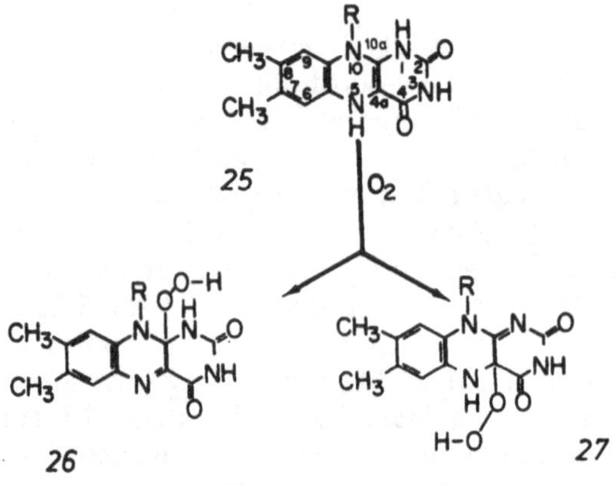

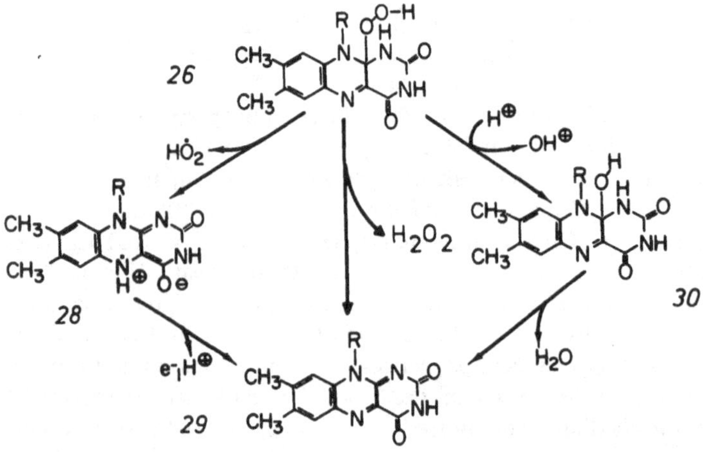

Scheme 7

a phenolic compound is an electrophilic reaction. The proposed oxenium ion possesses the chemical properties required. Recently, it was shown that p-hydroxybenzoate hydroxylase [218,219], and also luciferase [220], form a labile flavin-C(4a)-hydroperoxide. In the latter case the structure of the intermediate was elucidated by NMR [221]. During the catalytic action of these enzymes several intermediates (consecutive break-down products of (*27*)? absorbing between 370 and about 420 nm are formed [219]. Structure (*26*) was for a long time not accepted because the C(10a) position of flavin was believed to be chemically rather unreactive until it was found that σ-complexes are formed between quaternary flavinium salts and methoxide [38]. As shown later by NMR the formation of C(10a) σ-adducts is restricted to flavinium salts alkylated at N(1) [222,223]. In addition, careful oxidation of methanolic or acetonitrile solution of dimethyl-1,5-dihydroflavin yields a pure spectrum of the corresponding C(10a)-hydroperoxide [38]. The spectrum of this intermediate is very similar to that of the corresponding methoxide adduct. The introdution of more bulky alkyl substituents into position 10 of these model compounds yields visible absorption spectra of the σ-complexes resembling those of the intermediates observed during catalysis of the flavoprotein monooxygenases [224]. Similar studies were reported by Mager [225], who confirmed the data mentioned above.

Also other mechanisms of oxygen activation by flavohydroquinone have been postulated, among others the formation of oxaziridines, dioxetanes [226] and carbonyl oxides [227], but the resulting chromophores would probably absorb at wavelengths shorter than 370 nm. However, Bruice and his co-workers [61,228-231] have gathered overwhelming evidence for the existence of flavin-C(4a)-hydroperoxide and its capability to oxidize various compounds. These model reactions show the remarkable feature that flavin hydroperoxide behaves in many cases like a dioxygenase, i.e. transfer of hydroperoxide to the substrate [61]. It should however be kept in mind that such model studies, as impressive as they may be, cannot tell the whole story about flavoprotein-catalyzed oxygenation reactions. It is evident that more refined model studies and a better understanding of the enzyme-catalyzed reaction mechanism are needed to explain fully the experimental observations. In addition, no satifactory explanation has yet been offered for the fact that flavohydroquinone reacts rather smoothly with molecular oxygen, while it violates the law of spin conservation, i.e. reaction of a singlet with a triplet molecule. Nevertheless, the existence of flavin hydroperoxides cannot be denied anymore. It seems most logical that the hydroperoxide must be formed in the sequence *(27)*→*(28)* (Scheme 7) because C(4a) can easily accommodate the unpaired electron leading to the stable (protein-bound) neutral flavosemiquinone (Table 6). Concerning the importance of *(26)* and *(27)* in biological oxidation reactions (including monooxygenations) the proposed corresponding reaction path needs still further experimental support. However, in the opinion of the author *(26)* may well play an important role in monooxygenation reactions if the protein supports its formation. Recent ^{15}N-NMR results [232] on free and protein-bound flavins indicate that N(10) could assist the sequence *(26)*→*(30)* and the easy elimination of H_2O from *(30)* to yield *(29)* (Scheme 7). In this context the theoretical studies of Goddard and Harding [233], which show that both *(26)* and *(27)* are feasible intermediates, but that *(27)* is energetically more favoured, may provide a good basis for the design of further experiments.

5 Flavoproteins

5.1 Redox Shuttles

Flavoproteins occupy a key position among the oxidizing enzymes. No significant extent of biological oxidation occurs except through flavoproteins as part of electron transfer mechanism. Among the great number of flavoproteins, about 100, there are four which are as yet not believed to be involved in biological oxidations: Oxynitrilase which catalyzes the formation of dextrorotatory oxynitriles [23], glyoxylate carboligase which catalyzes the formation of hydroxymalonic semialdehyde [234] and CO_2 from glyoxylate and finally, riboflavin-binding protein from eggs which is believed to function as a storage for vitamin B_2 and as a riboflavin carrier [235]. The real biological function of old yellow enzyme which catalyzes the oxidation of NADPH is also not yet known [7].

Theoretically, flavoproteins could make use of all three redox states during catalysis (cf. Scheme 3). But it turned out that flavoproteins use selectively only certain redox states, depending on their biological functions. In the following the present state of knowledge will be summarized briefly in a simplified manner. Thus, simple flavoprotein oxidases, i.e. flavoproteins not containing a metal prosthetic group, shuttle between the flavoquinone and flavohydroquinone state forming hydroperoxide (4).

$$Fl_{red}H_3 + O_2 \rightarrow Fl_{ox}H + H_2O_2 \quad (4)$$

Examples of this class of enzymes are glucose oxidase and D-amino acid oxidase [236]. The classification of the flavoproteins used here is that originally suggested [34] which has been modified recently [60]. In the author's own view the original classification has the advantage of being simple and yet quite useful whereas the new classification does not add to simplify and classify the rather complex picture of flavoprotein catalysis. Nevertheless, in flavoprotein oxidases, the 1,5-dihydroflavin is very reactive towards O_2. On the other hand, the two-electron reduced form of flavoprotein oxidases reacts slowly with pure one-electron acceptors, e.g. ferricyanide. That the two-electron transition is biologically favoured in these enzymes explains why they can react easily with sulfite [34].

The oxygen reactivity of flavohydroquinone bound to apoflavoprotein dehydrogenases can vary considerably from fast (flavodoxins), moderate (xanthine oxidase) to nil (succinate dehydrogenase) [237]. Most, but not all, flavoprotein dehydrogenases contain one or more types of metal prosthetic groups, e.g. xanthine oxidase contains also Fe and Mo. Since these metal ions are involved in electron flux, their possible participation in the reaction with O_2 cannot be excluded. Much evidence, however, indicates that the flavin is involved in the one-electron reduction of O_2, as shown in Equation (5).

$$Fl_{red}H_3 + O_2 \rightarrow \dot{F}lH_2 + O_2H\,. \quad (5)$$

The simple flavoprotein dehydrogenases, i.e. those not containing a metal prosthetic group, such as flavodoxins, have been studied widely and found to react according to reaction (5). In vivo and in vitro, the flavoprotein dehydrogenases from the flavosemiquinone which is a catalytic intermediate. The stability of the flavosemiquinone differs greatly as far as in vitro reactions are concerned. *Azotobacter vinelandii* falvodoxin forms one of the most air-stable neutral flavosemiquinone [238]. In contrast to other dehydrogenases, the two-electron reduced flavodoxins react smoothly with O_2. This different reactivity towards O_2 is most probably related to the conformation of the prosthetic group, i.e. planar *vs.* bent (cf. below). The two-electron reduced flavoprotein dehydrogenases are in general, with the exception of flavodoxins, more reactive towards ferricyanide than O_2.

The transition flavoquinone-flavosemiquinone seems not to be useful in flavoproteins catalysis. Only trimethylamine dehydrogenase electron-acceptor flavoprotein, isolated from bacterium W3A1 [239], makes probably use of this shuttle [240,241]. The enzyme forms a very air-stable anionic flavosemiquinone.

Although flavoprotein monooxygenases form a separate class, formally they can be considered to react according to Equation (4) as far as the redox states involved in catalysis are concerned.

Besides flavocoenzymes, also other prosthetic groups or coenzymes can be involved in biological oxidation reactions, i.e. metal ions. Whenever e.g. electrons must be transported from reduced pyridine nucleotides to metal-containing enzymes, which are one-electron acceptors or donators, then a flavoprotein is required as mediator. In such cases the flavoprotein receives two electrons and one proton (hydride-transfer?) from NAD(P)H yielding the flavohydroquinone state of the protein which in turn splits the electron pair into two electron equivalents which can be accommodated by the metal-containing enzyme. The reverse reaction, i.e. uptake of two consecutive one-electron equivalents and donation of a electron pair, if of course also possible, for instance in xanthine dehydrogenase [242]. In these reactions flavoproteins make use of all three redox states. These statements need a cautious note. The donation or acceptance of two one-electron equivalents is only feasible when the electrons are of equipotential energy. However, the redox potential between flavoquinone and flavosemiquinone, on the one hand, and that between flavosemiquinone and flavohydroquinone, on the other hand, can differ considerably. Since only a few data are available, the feasibility of these reaction sequences must still await experimental proof.

The latter problem has been investigated in detail in two complex flavoprotein systems, i.e. NADPH-sulfite reductase [243,244] and microsomal NADPH-cytochrome P-450 reductase [245]. These enzymes are complex in several regards. They contain iron part of which may be associated with heme or heme-like chromophores. Even more interesting (at least in context with the subject concerned here) is the fact that these enzymes need FMN *and* FAD as prosthetic groups, in contrast to the above mentioned "simple" flavoproteins containing only one particular flavocoenzyme, not considering the variation in stoichiometry in different enzymes. Anyhow, NADPH-sulfite reductase contains 4 FAD and 4 FMN per eight identical protein subunits whereas NADPH-cytochrome P-450 reductase contains 1 FAD and 1 FMN per polypeptide chain. Although the electron flux in NADPH-sulfite reductase exhibits a much more complex reaction sequence than NADPH-cytochrome P-450 reductase, the common feature of both enzymes will be presented in a simplified approach. It could be demonstrated convincingly that the entry port of electrons is FAD shuttling between the flavoquinone and flavohydroquinone state during catalysis, but making use of the intermediate flavosemiquinone level. FMN, on the other hand, forms a rather stable semiquinone and shuttles between the semiquinone and the hydroquinone state. The overall reactions are presented by Eqs. (6) to (8).

$$\mathrm{NADPH} + \mathrm{FAD} \rightarrow \mathrm{NADP}^+ + \mathrm{FADH}_2\,, \tag{6}$$

$$\mathrm{FADH}_2 + 2\,\dot{\mathrm{F}}\mathrm{MNH} \rightarrow \mathrm{FAD} + 2\,\mathrm{FMNH}_2\,, \tag{7}$$

$$2\,\mathrm{FMNH}_2 + 2\,\mathrm{A(acceptor)} \rightarrow 2\,\dot{\mathrm{F}}\mathrm{MNH} + 2\,\mathrm{A}^-\,. \tag{8}$$

To transport the two electrons from NADPH to the acceptor molecule (A), the one-electron transfer reactions must proceed in two consecutive steps. These two enzymes demonstrate how nature is making use of one and the same redox system to split the incoming electron-pair into single electrons of equipotential energy to reduce a particular acceptor system.

Table 7. Some Redox Potentials (E'_o in mV) of Free and Protein-Bound Flavins

Enzyme	Prosthetic group	pH	Redox potentials				
			$Fl_{ox}H \rightleftarrows \dot{F}lH_2$ E_2	$\dot{F}lH_2 \rightleftarrows Fl_{red}H_3$ E_1	$Fl_{ox}H \rightleftarrows Fl_{red}H_3$ E'_0	$\Delta E_{1,2}$	Ref.
—	Rfl	6.95	−231	−167	−199	− 64	146)
	FMN	7.00	−238	−172	−205	− 66	246)
Clostridium MP flavodoxin	FMN	7.00	− 92	−399	−245	+307	247)
M. elsdenii flavodoxin	FMN	7.00	−115	−372	−243	+257	248)
A. vinelandii flavodoxin	FMN	8.2	+ 50	−495	−272	+445	249)
C. pasteurianum flavodoxin	FMN	7.0	−132	−419	−275	+287	247)
D. vulgaris flavodoxin	FMN	7.8	−149	−438	−293	+289	250)
E. coli flavodoxin	FMN	7.7	−240	−410	−325	+170	251)
A. nidulans flavodoxin	FMN	7.0	−180	−470	−325	+290	252)
S. lividus flavodoxin	FMN	7.0	− 50	−450	−250	+400	253)
Old yellow enzyme	FMN	7.0	—	—	−123	—	254)
Glucose oxidase	FAD	5.3	− 63	− 65	− 64	+ 2	159)
		9.3	−200	−240	−220	+ 40	159)
D-amino acid oxidase	FAD	7.0	—	—	− 4	—	255)
Thiamine dehydrogenase	FAD	7.2	+ 80	+ 30	+ 55	− 50	256)
Xanthine dehydrogenase	FAD	7.8	−345	−377	−361	+ 32	242)
		6.8	−294	−330	−312	+ 36	242)
Microsomal NADPH-cytochrome P-450	FAD	7.0	−110	−270	−190	+160	257)
reductase	FMN	7.0	−290	−365	−328	+ 75	257)
NADPH-sulfite reductase	FAD	7.0	−322	−382	−352	+ 60	243, 258)
	FMN	7.0	−327	−152	−241	−175	243, 258)
Adrenodoxin reductase	FAD	7.0	−320	−200	−260	−120	259)
Electron-transferring flavoprotein ubiquinone oxido reductase	FAD	7.0	—	—	− 45	—	260)
NADH-cytochrome b_5 reductase	FAD	7.0	—	—	−258	—	261)
L-Amino acid oxidase	FAD	7.0	—	—	+ 33	—	262)
	FAD a)	7.0	—	—	−200	—	262)
p-Hydroxybenzoate hydroxylase from *Pseudomonas fluorescens*	FAD b)	7.0	−160	−265	−212	+105	263)
A. vinelandii transhydrogenase	FAD b)	7.0	—	—	−280	—	263)
O_2/O_2^-		7.0	−330	—	—	—	264)
O_2/H_2O_2		7.0	—	—	+270	—	265)

a) Low-temperature inactivated enzyme, FAD is still bound; b) Preliminary results

Flavoproteins are redox systems. It could, therefore, be expected that their redox potentials would tell us something about their function and eventually about the possible reaction mechanism. In Table 7 an overview of available data is given. Although it seems too early to draw some definite conclusions from these results, a few points are worth mentioning. All flavodoxins, i.e. pure one-electron transferring flavoproteins, show a large difference between the E_1 and E_2 potentials amounting from about 200 to 450 mV. These values should be compared to that of free flavin (about 60 mV). The corresponding values of the flavoprotein oxidases, show, in contrast, only a relative small difference, i.e. glucose oxidase, thiamine dehydrogenase (forms H_2O_2), xanthine dehydrogenase (reduces NAD^+). These enzymes are two-electron reducing enzymes with respect to the terminal electron acceptors and the E_0' values are in the range needed to carry out the corresponding reactions. In addition, where the terminal acceptor is molecular oxygen the E_0' values are relatively positive. On the other hand, the proposed FAD involved in the oxidation of NADPH ($E_0' = -320$ mV) in NADPH-cytochrome P-450 reductase possesses the expected E_0' value whereas that of NADPH-sulfite reductase seems too low to accept the hydride-equivalent easily from NADPH. Therefore, these values must be used with caution, also considering the complexity of these systems. The flavoprotein transhydrogenase from *A. vinelandii* reacts only with pyridine nucleotides. It is interesting that this enzyme exhibits the expected behaviour showing only a two-electron transition. On the other hand, p-hydroxybenzoate hydroxylase gives both a E_1 and E_2, while it does not form in general any intermediate semiquinone, either chemically or catalytically. Its $\Delta E_{1,2}$ value resembles that of the two complex enzyme systems. This could be taken as an indication that the reaction mechanism of this enzyme is even more complex than believed and could fit the theory developed by Conrad [266]. From Table 7 we also learn that the properties of the flavocoenzymes are taylored by the apoproteins to fit a particular biological reaction as was suggested earlier [40]. A good illustration of this fact is the finding that the redox potential of L-amino acid oxidase changes drastically upon reversible low-temperature inactivation [262] (cf. Table 7).

5.2 Structural Studies

During the past decade many papers have appeared with the optimistic title "On the mechanism of . . .". It has become, however, clear that the mechanisms governing the flavoprotein-catalyzed reactions are still not well understood, though different approaches in the study of these enzymes have opened new perspectives. The wealth of data available must be complemented by physical studies yielding information on the dynamic structure of the biomolecules on the molecular and submolecular level. The knowledge of the three-dimensional structure of an enzyme is a desirable basis to interpret results obtained by other physical techniques. The flavodoxins are probably the most intensively studied flavoproteins. Therefore, a short discussion is devoted to these proteins in the following. The three-dimensional structures of a few flavoproteins have been elucidated, others are currently under active investigation [33,39,45–49]. The most thoroughly studied protein is flavodoxin from *Clostridium MP* [267,268]. All three redox states were investigated. The three-dimensional data show that FMN possesses a planar structure in all three redox states, though the flavohydroquinone state deviates from coplanarity by a few degrees. One of the most surprising findings to most of the researchers in the field was the fact that the apolar part of FMN, i.e. the benzene subnucleus, is on the surface of the protein and in contact with bulk solvent, whereas the pyrimidine nucleus of FMN is burried in the apoprotein. If we look at Table 7, we may understand that this arrangement is the only way to modulate the redox potential of flavoproteins by specific interactions (e.g. hydrogen bonds) with amino acid residues of the apoprotein. Indeed, interactions are formed at N(1), C(2α), N(3), C(4α) and N(5) of FMN. The results were recently reviewed in detail by Mayhew and Ludwig [63,269]. From potentiometric studies on the closely related flavodoxin from *Megasphaera elsdenii* it has been suggested that the flavohydroquinone is in the anionic state [270]. This was recently ascertained by NMR [271]. The same holds for *A. vinelandii* flavodoxin and the other flavodoxins. The fact that anionic flavohydroquinone in flavodoxins is forced to acquire a planar structure makes the flavodoxins powerful reducing agents. It was suggested that the electron-transfer of reduced flavodoxins occurs *via* the $C(8)CH_3$ group of the prosthetic group [272]. This suggestion was deduced from the fact that both methyl groups of protein-bound FMN are in contact with bulk solvent [267,268]. This suggestion must be dismissed on the following grounds: 1) Replacement of FMN in *M. elsdenii* flavodoxin by a FMN derivative, where the benzene nucleus is fully reduced, still yields a partly catalytically active protein, although electron flow *via* $C(8)CH_3$ is impossible [273]. 2) In most flavosemiquinones [274,275] and flavoquinones in flavodoxins [276] water and free amino acids have access to the N(5)-C(4a) part of the prosthetic group. 3) Electron transfer between the semiquinone and hydroquinone in flavodoxin is fast whereas that between the semiquinone and quinone is slow [277]. 4) ^{15}N-NMR data prove that N(5) of bound FMN is the entry and exit port of electrons. In p-Hydroxybenzoate hydroxylase the methyl groups of the benzene subnucleus of FAD are also exposed to solvent but catalysis occurs via N(5)-C(4a) of bound FAD [45,46].

The three-dimensional structures, or part of it, are also known for *Desulfovibrio vulgaris* and *Anacystis nidulans* flavodoxins. These results, including those obtained on *C. MP.*, were recently summarized by Adman [278]. Hence, these results will be discussed only briefly. The x-ray structures show that the isoalloxazine ring is embedded in a hydrophobic pocket of the apoprotein, i.e. flanked by at least one aromatic amino acid residue. During the redox transitions, especially from the oxidized to the semiquinone state, small conformational changes occur and contacts with the isoalloxazine ring are formed or broken. These conformational transitions form probably a kinetic barrier so that the semiquinone state is "trapped" by the apoprotein and, therefore, rather stable towards oxidation by molecular oxygen.

Until recently, it was not known that flavoproteins contain also covalently bound phosphate. *A. vinelandii* flavodoxin, L-amino acid oxidase, glucose oxidase, NADPH-cytochrome c reductase, thiamine dehydrogenase possess one mole of covalently bound phosphate/per mole protein [53,279]. The function of these phosphate residues is not yet known. In glucose oxidase the phosphate residue is linked to two amino acid residues implicating that it functions like a disulfide bridge [279].

Besides the x-ray and NMR technique, two other valuable techniques should be mentioned. Raman spectroscopy has been applied to a few flavoproteins. This technique should be able to reveal the interaction sites between the isoalloxazine ring and amino acid residues [280-283]. Published data reveal that the complexity of the Raman spectra makes it a difficult task to extract the desired information. This is partly due to the fact that the observed lines cannot yet be definetely assigned, although some progress has been booked recently in this respect [284]. However, it can be expected that the combination of NMR and Raman spectroscopy will make it possible to reveal the subtle interactions between prosthetic group and apoflavoprotein. These interactions are believed to be the key for a profound understanding of the different functions of flavoproteins.

The other technique of great potential for obtaining structure information is the laser-induced luminescence. Not all, but many flavoproteins exhibit a weak to a strong fluorescence emission. Time-dependent anisotropy studies on pig heart lipoamide dehydrogenase [285] and the pyruvate dehydrogenase multi-enzyme complex [286], where lipoamide dehydrogenase is one of the constituents, revealed that the prosthetic group, i.e. the isoalloxazine ring, possesses internal flexibility. This observation is unexpected and, at a first glance, in contradiction to the concept of specific interactions between the prosthetic group and apoprotein. This apparent discrepancy is a result of thinking in static structures, as imposed by x-ray studies, rather than dynamic structures, which are closer to reality. Specific interactions must play an important role in the catalysis of flavoproteins, but these interactions can still be of dynamic nature. It is now generally accepted that protein molecules can undergo local and molecular structural fluctuations [278].

Finally, it should be mentioned that theoretical studies are needed to support or reject certain hypotheses deduced from experimental studies, and, of course, *vice versa*. In the field of flavin and flavoprotein research a few theoretical papers were published which are useful guides for the experimentalists [55,287,288].

6 Concluding Remarks

It was attempted to describe the fundamental chemical properties of flavins in some depth and to correlate them, whenever possible, to those of flavoproteins. It is obvious that, considering the wealth of information available, not all properties of flavin received the same attention which is reflected in the literature cited.

Instead of reviewing the whole field the new and possible future developments received first priority. For instance, the photochemical data on flavins were only mentioned incidentally because, in the author's view, the photochemical reactivity of flavin can be related less directly to the catalytic properties of most flavoproteins. However, for the reader interested in this subject the few references given should make possible an easy access to the appropriate literature.

I am grateful to Mrs. J. C. Toppenberg-Fang, Mr. B. J. Sachteleben, Mr. M. M. Bouwmans, Dr. C. van Dijk, Ir. C. T. W. Moonen for their assistance in the preparation of the manuscript. I am indebted to Dr. C. van Dijk for the allowance to use unpublished results.

The unpublished work cited in this paper has been carried out under the auspices of the Netherlands Foundation for Chemical Research (S.O.N.) with financial aid from the Netherlands Organization for the Advancement of Pure Research (Z.W.O.).

7 References

1. Blyth, A. W.: J. Chem. Soc. *35*, 530 (1879)
2. Rosenthaler, L.: Biochem. Z. *14*, 235 (1908)
3. Dixon, M., Kodama, K.: Biochem. J. *20*, 1104 (1926)
4. Warburg, O., Christian, W.: Biochem. Z. *254*, 438 (1932)
5. Kuhn, R., Reinemund, K., Weygand, F.: Ber. Dtsch. Chem. Ges. *67*, 1460 (1934)
6. Karrer, P., Schöpp, K., Benz, F.: Helv. Chim. Acta *18*, 426 (1935)
7. Theorell, H.: Biochem. Z. *278*, 263 (1935)
8. Kuhn, R., Rudy, H. Weygand, F.: Ber. Dtsch. Chem. Ges. *69*, 1543 (1936)
9. Warburg, O. Christian, W.: Biochem. Z. *296*, 294 (1938)
10. Straub, F. B.: Biochem. J. *33*, 787 (1939)
11. Weber, G.: Biochem. J. *47*, 114 (1950)
12. Sanadi, D. R., Littlefield, J. W., Bock, R. M.: J. Biol. Chem. *197*, 851 (1952)
13. Green, D. E., Beinert, H.: Biochem. Biophys. Acta *11*, 599 (1953)
14. Christie, S. M. H., Kenner, G. W., Todd, A. R.: J. Chem. Soc. *1954*, 46
15. Sutton, W. B.: J. Biol. Chem. *210*, 309 (1954)
16. Kearney, E. B., Singer, T. P.: Biochem. Biophys. Acta *17*, 596 (1955)
17. Boeri, E., Tosi, L.: Arch. Biochem. Biophys. *60*, 463 (1956)
18. Beinert, H.: J. Amer. Chem. Soc. *78*, 5323 (1956)
19. Steyn-Parvé, E. P., Beinert, H.: J. Biol. Chem. *233*, 853 (1958)
20. Lang, C. A., Nason, A.: J. Biol. Chem. *234*, 1874 (1959)
21. Kearney, E. B.: J. Biol. Chem. *235*, 865 (1960)
22. Katagiri, M., Yamamoto, S., Hayaishi, O.: J. Biol. Chem. *237*, 2413 (1962)
23. Becker, W. et al.: Biochem. Z. *337*, 156 (1963)
24. Hemmerich, P.: Helv. Chim. Acta *47*, 464 (1964)
25. McCormick, D. B., Chassy, B. M., Tsibris, J. C. M.: Biochem. Biophys. Acta *89*, 447 (1964)
26. Mager, H. I. X., Berends, W.: Biochem. Biophys. Acta *118*, 440 (1966)
27. Massey, V., Palmer, G.: Biochemistry *5*, 3181 (1966)
28. Kierkegaard, P. et al.: J. Chem. Soc. Chem. Commun. *1967*, 288
29. Siegel, L. M., Monty, K. J., Kamin, H.: Federaton Proc. *26*, 858 (1967)
30. Ehrenberg, A., Müller, F., Hemmerich, P.: Eur. J. Biochem. *2*, 286 (1967)
31. Machinist, J. M., Dehner, E. W., Ziegler, D. M.: Arch. Biochem. Biophys. *125*, 858 (1968)
32. Hemmerich, P. et al,: FEBS Lett. *3*, 37 (1969)
33. Ludwig, M. L. et al.: J. Biol. Chem. *244*, 6047 (1969)
34. Massey, V. et al.: J. Biol. Chem. *244*, 3999 (1969)
35. Müller, F., Massey, V.: J. Biol. Chem. *244*, 4007 (1969)
36. Monteilhet, C., Risler, J. L.: Eur. J. Biochem. *12*, 165 (1970)
37. Tanaka, M. et al.: Biochemistry *10*, 3041 (1971)
38. Müller, F., in: Flavins and Flavoproteins. Kamin, H. (ed.), p. 363. Baltimore: University Park Press 1971
39. Watenpaugh, K. D. et al.: Proc. Natl. Acad. Sci. USA *69*, 3185 (1972)
40. Müller, F.: Z. Naturforsch. *27b*, 1023 (1972)
41. Ghisla, S., Mayhew, S. G.: J. Biol. Chem. *248*, 6568 (1973)
42. Miura, R. et al.: J. Chem. Soc. Chem. Commun. *1973*, 703
43. Porter, D., Voet, J., Bright, H.: J. Biol. Chem. *248*, 440 (1973)
44. Ghisla, S. et al.: Biochemistry *13*, 589 (1974)
45. Drenth, J., Hol, W. G. J., Wierenga, R. K.: J. Biol. Chem. *250*, 5268 (1975)
46. Wierenga, R. K. et al.: J. Mol. Biol. *131*, 55 (1979)
47. Schulz, G. E. et al.: FEBS Lett. *54*, 86 (1975)
48. Schulz, G. E. et al.: Nature *273*, 120 (1978)
49. Smith, W. W. et al.: J. Mol. Biol. *94*, 123 (1975)
50. Shinkai, S., Sakuma, Y., Yoneda, F.: J. Chem. Soc. Chem. Commun. *1976*, 986
51. Moore, E. G., Cardemil, E., Massey, V.: J. Biol. Chem. *253*, 6413 (1978)
52. Eirich, L. D., Vogels, G. D., Wolfe, R. S.: Biochemistry *17*, 4583 (1978)
53. Edmondson, D. E., James, T. L.: Proc. Natl. Acad. Sci. USA *76*, 3786 (1979)

54. Eweg, J. K. et al.: J. Amer. Chem. Soc. *102*, 51 (1980)
55. Palmer, M. H., Simpson, I., Platenkamp, R. J.: J. Mol. Structure *66*, 243 (1980)
56. Beinert, H.: The Enzymes, 2nd edit., Vol. 2, p. 339, New York: Academic Press 1960
57. Michaelis, L., Schubert, M. P., Smythe, C. V.: J. Biol. Chem. *116*, 587 (1936)
58. Michaelis, L., Schwarzerbach, G.: J. Biol. Chem. *123*, 527 (1938)
59. Hemmerich, P., Fallab, S.: Helv. Chim. Acta *41*, 498 (1958)
60. Massey, V., Hemmerich, P.: Biochem. Soc. Trans. *8*, 246 (1980)
61. Bruice, T. C.: Acc. Chem. Res. *13*, 256 (1980)
62. Walsh, C.: Acc. Chem. Res. *13*, 148 (1980)
63. Simondson, R. P., Tollin, G.: Mol. Cell. Biochemistry *33*, 13 (1980)
64. Hemmerich, P.: Prog. Chem. Org. Nat. Prod. *33*, 451 (1976)
65. Slater, E. C. (ed.): Flavins and Flavoproteins. Amsterdam: Elsevier 1966
66. Yagi, K. (ed.): Flavins and Flavoproteins. Tokyo: University of Tokyo Press 1968
67. Kamin, H. (ed.): Flavins and Flavoproteins. Baltimore: University Park Press 1971
68. Yagi, K. (ed.): Reactivity of Flavins. Tokyo: University of Tokyo Press 1975
69. Singer, T. P. (ed.): Flavins and Flavoproteins. Amsterdam: Elsevier 1976
70. Ostrowski, W. (ed.): Flavins and Flavoproteins. Warsaw and Cracow: Polish Scientific Publishers 1977
71. Yagi, K., Yamano, T. (eds.): Flavins and Flavoproteins. Tokyo: Japan Scientific Societies Press 1980
72. Massey, V., Williams, C. H. (eds.): Flavins and Flavoproteins. New York: Elsevier 1981, to be published
73. Kuhn, R., Weygand, F.: Ber. Dtsch. Chem. Ges. *67*, 1409 (1934)
74. Plant, G. W. E., Smith, C. M.: Ann. Rev. Biochem. *43*, 899 (1974)
75. Mailänder, B., Bacher, A.: J. Biol. Chem. *251*, 3623 (1976)
76. Walsh, C. et al.: Biochemistry *17*, 1942 (1978)
77. Arsenis, C., McCormick, D. B.: Biochem. Biophys. Acta *92*, 440 (1964)
78. Edmondson, D. E., Tollin, G.: Biochemistry *10*, 124 (1971)
79. Mayhew, S. G. et al.: Eur. J. Biochem. *44*, 570 (1974)
80. Ghisla, S., Mayhew, S. G.: Eur. J. Biochem. *63*, 373 (1976)
81. Kuhn, R., Weygand, F.: Ber. Dtsch. Chem. Ges. *68*, 1282 (1935)
82. Karrer, P., Schöpp, K., Benz, F.: Helv. Chim. Acta *18*, 1435 (1935)
83. Tishler, M. et al.: J. Amer. Chem. Soc. *69*, 1487 (1947)
84. Müller, F., Hemmerich, P.: Helv. Chim. Acta *49*, 2352 (1966)
85. Scola-Nagelschneider, G., Hemmerich, P.: Eur. J. Biochem. *66*, 567 (1976)
86. Moffatt, J. G., Khorana, H. G.: J. Amer. Chem. Soc. *83*, 649 (1961)
87. Roseman, S. et al.: J. Amer. Chem. Soc. *83*, 659 (1961)
88. Cramer, F., Neunhoeffer, H.: Ber. Dtsch. Chem. Ges. *95*, 1664 (1962)
89. Mayhew, S. G., Strating, M. J. J.: Eur. J. Biochem. *59*, 539 (1975)
90. Massey, V., Mendelsohn, L. D.: Anal. Biochem. *95*, 156 (1979)
91. Theorell, H.: Biochem. Z. *272*, 155 (1934)
92. Warburg, O., Christian, W.: Biochem. Z. *298*, 150 (1938)
93. Green, D. E., MII, S., Kohout, P. M.: J. Biol. Chem. *217*, 551 (1955)
94. Chi, T. F. et al.: Scientia Sinica *14*, 1193 (1965)
95. Salach, J. et al.: Eur. J. Biochem. *26*, 267 (1972)
96. Singer, T. P., Edmondson, D. E.: FEBS Lett. *42*, 1 (1974)
97. Edmondson, D. E., Singer, T. P.: FEBS Lett. *64*, 255 (1976)
98. Hamm, H.-H., Decker, K.: Arch. Microbiol. *119*, 65 (1978)
99. Bullock, F. I., Jardetzky, O.: J. Org. Chem. *30*, 2056 (1965)
100. Hemmerich, P., Prijs, B., Erlenmeyer, H.: Helv. Chim. Acta *42*, 2164 (1959)
101. Smith, S. B., Brüstlein, M., Bruice, T. C.: J. Amer. Chem. Soc. *96*, 3696 (1974)
102. Ghisla, S., Hartmann, U., Hemmerich, P.: Angew. Chem. *82*, 669 (1970)
103. Knappe, W. R.: Justus Liebigs Ann. Chem. *1979*, 1062
104. Ghisla, S. et al.: Biochemistry *19*, 2537 (1980)
105. Hayashi, S., Nakamura, S., Suzuki, M.: Biochem. Biophys. Res. Commun. *96*, 924 (1980)
106. Frisell, W. R., MacKenzie, C. G.: J. Biol. Chem. *237*, 94 (1962)
107. Kenney, W. C. et al.: J. Biol. Chem. *254*, 4689 (1979)

108. Edmondson, D. E., Kenney, W. C.: Biochem. Biophys. Res. Commun. *68*, 242 (1976)
109. Kenney, W. C. et al.: Biochemistry *15*, 4931 (1976)
110. Kenney, W. C. et al.: Biochem. Biophys. Res. Commun. *71*, 1194 (1976)
111. Kenney, W. C. et al.: FEBS Lett. *97*, 40 (1979)
112. Möhler, H., Brühmüller, M., Decker, K.: Eur. J. Biochem. *29*, 152 (1972)
113. Pinto, J. T., Frisell, W. R.: Arch. Biochem. Biophys. *169*, 483 (1975)
114. Walker, W. H. et al.: Eur. J. Biochem. *26*, 279 (1972)
115. Kenney, W. C., Kröger, A.: FEBS Lett. *73*, 239 (1977)
116. Ohishi, N., Yagi, K.: Biochem. Biophys. Res. Commun. *86*, 1084 (1979)
117. McIntire, W. et al.: J. Biol. Chem. *255*, 6553 (1980)
118. Kearney, E. B. et al.: Eur. J. Biochem. *24*, 321 (1971)
119. Kenney, W. C., Singer, T. P.: J. Biol. Chem. *252*, 4767 (1977)
120. Kenney, W. C., McIntire, W., Yamanaka, R.: Biochem. Biophys. Acta *483*, 467 (1977)
121. Steenkamp, D. J., Kenney, W. C., Singer, T. P.: J. Biol. Chem. *253*, 2812 (1978)
122. Steenkamp, D. J., McIntire, W., Kenney, W. C.: J. Biol. Chem. *253*, 2818 (1978)
123. Singer, T. P., Edmondson, D. E.: Methods Enzymol. *66*, 253 (1980)
124. Mayhew, S. G., Massey, V.: Biochem. Biophys. Acta *235*, 303 (1971)
125. Whitfield, C. D., Mayhew, S. G.: J. Biol. Chem. *249*, 2810 (1974)
126. Ghisla, S., Mayhew, S. G.: Eur. J. Biochem. *63*, 373 (1976)
127. Whitfield, C. D., Mayhew, S. G.: J. Biol. Chem. *249*, 2811 (1974)
128. Mayhew, S. G. et al.: Eur. J. Biochem. *44*, 579 (1974)
129. Ghisla, S., Mayhew, S. G.: Methods Enzymol. *66*, 241 (1980)
130. Schöllnhammer, G., Hemmerich, P.: Eur. J. Biochem. *44*, 561 (1974)
131. Otani, S. et al.: J. Antibiot. *27*, 88 (1974)
132. Kasai, S., Miura, R., Matsui, K.: Bull. Chem. Soc. Japan *48*, 2877 (1975)
133. Otani, S., Kasai, S., Matsui, K.: Methods Enzymol. *66*, 235 (1980)
134. Tachibana, S.: J. Vitaminol. *18*, 210 (1972)
135. Tachibana, S., Murakami, T.: J. Nutr. Sci. Vitaminol. *21*, 61 (1975)
136. Cheesman, P., Tomms-Wood, A., Wolfe, R. S.: J. Bacteriol. *112*, 527 (1972)
137. Ashton, W. et al.: J. Amer. Chem. Soc. *101*, 4419 (1979)
138. Visser, C. M.: Naturwissenschaften *67*, 549 (1980)
139. Dudley, K. H. et al.: Helv. Chim. Acta *47*, 1354 (1964)
140. Visser, A. J. W. G., Müller, F.: Helv. Chim. Acta *62*, 593 (1979)
141. Harbury, H. A. et al.: Proc. Natl. Acad. Sci. USA *45*, 1708 (1959)
142. Eweg, J. K. et al.: J. Amer. Chem. Soc. *103* (1981), submitted
143. Visser, A. J. W. G., Van Hoek, A., Müller, F.: Helv. Chim. Acta *63*, 1296 (1980)
144. Wahl, Ph. et al.: FEBS Lett. *44*, 67 (1974)
145. Wassink, J. H., Mayhew, S. G.: Anal. Biochem. *68*, 609 (1975)
146. Draper, R. D., Ingraham, L. L.: Arch. Biochem. Biophys. *125*, 802 (1968)
147. Land, E. J., Swallow, A. J.: Biochemistry *8*, 2117 (1969)
148. Gibson, Q. H., Massey, V., Atherton, N. M.: Biochem. J. *85*, 369 (1962)
149. Kuhn, R., Ströbele, R.: Chem. Ber. Dtsch. Ges. *70*, 753 (1937)
150. Müller, F. et al.: Eur. J. Biochem. *25*, 573 (1972)
151. Müller, F., Hemmerich, P., Ehrenberg, A.: Eur. J. Biochem. *5*, 158 (1968)
152. Müller, F. et al.: Eur. J. Biochem. *14*, 185 (1970)
153. Ghisla, S. et al.: Liebigs Ann. Chem. *1973*, 1388
154. Ghisla, S. et al.: Biochemistry *13*, 589 (1974)
155. Kierkegaard, P. et al., in: Flavins and Flavoproteins, Kamin, H. (ed.), p. 1. Baltimore: University Park Press 1971
156. Visser, A. J. W. G. et al.: Eur. J. Biochem. *101*, 13 (1979)
157. Schuman, M., Massey, V.: Biochem. Biophys. Acta *227*, 500 (1971)
158. Massey, V., Ganther, H.: Biochemistry *4*, 1161 (1965)
159. Stankovich, M. T., Schopfer, L. M., Massey, V.: J. Biol. Chem. *253*, 4971 (1978)
160. Massey, V., Stankovich, M. T., Hemmerich, P.: Biochemistry *17*, 1 (1978)
161. Kuhn, R., Wagner-Jauregg, Th.: Ber. Dtsch. Chem. Ges. *66*, 1577 (1933)
162. Kuhn, R., Reinemund, K., Weygand, F.: Ber. Dtsch. Chem. Ges. *67*, 1460 (1934)
163. Warburg, O., Christian, W.: Biochem. Z. *263*, 228 (1933)

164. Smith, S. B., Bruice, T. C.: J. Amer. Chem. Soc. *97*, 2875 (1975)
165. Koziol, J., Hemmerich, P.: Liebigs. Ann. Chem. *1976*, 1276
166. Koziol, J., Tyrakowska, B., Müller, F.: Helv. Chem. Acta *64*, 1812 (1981)
167. Müller, F., Walker, W. H., Hemmerich, P.: Helv. Chim. Acta *49*, 2365 (1966)
168. Hemmerich, P.: Helv. Chim. Acta *43*, 1942 (1960)
169. Hemmerich, P., Prijs, B., Erlenmeyer, H.: Helv. Chim. Acta *42*, 2164 (1959)
170. Polyakova, N. A. et al.: Zh. Obsch. Khim. *42*, 465 (1972)
171. Moore, E. G., Ghisla, S., Massey, V.: J. Biol. Chem. *254*, 8173 (1979)
172. Massey, V., Ghisla, S., Moore, E. G.: J. Biol. Chem. *254*, 9640 (1979)
173. Porter, D. J. T., Voet, J. G., Bright, H. J.: Z. Naturforsch. *27b*, 1052 (1972)
174. Porter, D. J. T., Bright, H. J.: J. Biol. Chem. *252*, 4361 (1977)
175. Walker, W. H., Hemmerich, P., Massey, V.: Eur. J. Biochem. *13*, 258 (1970)
176. Knappe, W. R., Hemmerich, P., Justus Liebigs Ann. Chem. *1976*, 2037
177. Haas, W., Hemmerich. P.: Biochem. J. *181*, 95 (1979)
178. Turro, N. J.: Modern Molecular Photochemistry. Menlo Park: Benjamin-Cummings Publ. Co. 1978
179. Williams, R. F., Bruice, C. T.: J. Amer. Soc. *98*, 7752 (1976)
180. Novak, M., Bruice, T. C.: J. Amer. Chem. Soc. *99*, 8079 (1977)
181. Abeles, R., Maycock, A.: Acc. Chem. Res. *9*, 313 (1976)
182. Maycock, A. et al.: Biochemistry *15*, 114 (1976)
183. Ghisla, S. et al.: Biochemistry *15*, 1798 (1976)
184. Ghisla, S., Massey, V.: Biochemistry *18*, 4733 (1979)
185. Shinkai, S., Ando, R., Kunitake, T.: Biopolymers *17*, 2757 (1978)
186. Shinkai, S., Kamashita, T., Manabe, O.: J. Chem. Soc. Chem. Commun. *1979*, 301
187. Shinkai, S. in: Flavins and Flavoproteins. Yagi, K., Yamano, T. (eds.), p. 45. Tokyo: Japan Scientific Societies Press 1980
188. Hemmerich, P., Müller, F., Ehrenberg, A. in: Oxydases and Related Redox Systems. King, T. E., Mason, H. S., Morrison, M. (eds.), p. 157. New York: John Wiley and Sons 1965
189. Lauterwein, J., Hemmerich, P., Lhoste, J.-M.: Inorg. Chem. *14*, 2152 (1975)
190. Lauterwein, J., Hemmerich, P., Lhoste, J.-M.: Inorg. Chem. *14*, 2161 (1975)
191. Massey, V., Ghisla, S.: Ann. N.Y. Acad. Sci. *227*, 446 (1974)
192. Blankenhorn, G.: Eur. J. Biochem. *67*, 67 (1976)
193. Hemmerich, P. et al., in: Oxydases and Related Redox Systems. King, T. E., Mason, H. S., Morrison, M. (eds.), p. 3. Baltimore: University Park Press 1973
194. Porter, D. J. T., Bright, H. J., Voet, D.: Nature *269*, 213 (1977)
195. Voet, D., Rich, A.: Proc. Natl. Acad. Sci. USA *68*, 1151 (1971)
196. Inoue, M., Shibata, M., Ishida, T.: Biochem. Biophys. Res. Commun. *93*, 415 (1980)
197. Bear, C. A., Waters, J. M., Waters, T. N.: J. Chem. Soc. Chem. Commun. *1970*, 702
198. Langhoff, C. A., Fritchie, C. J.: J. Chem. Soc. Chem. Commun. *1970*, 20
199. Müller, F., Eriksson, L. E. G., Ehrenberg, A.: Eur. J. Biochem. *22*, 93 (1970)
200. Müller, F. et al.: Eur. J. Biochem. *116*, 17 (1981)
201. Eriksson, L. E. G., Hyde, J. S., Ehrenberg, A.: Biochem. Biophys. Acta *192*, 211 (1969)
202. Eriksson, L. E. G., Ehrenberg, A.: Biochem. Biophys. Acta *293*, 37 (1973)
203. Walker, W. H., Ehrenberg, A., Lhoste, J.-M.: Biochem. Biophys. Acta *215*, 166 (1970)
204. Favaudon, V., Lhoste, J.-M.: Biochemistry *14*, 4731 (1975)
205. Favaudon, V., Lhoste, J.-M.: Biochemistry *14*, 4739 (1975)
206. Barman, B. G., Tollin, G.: Biochemistry *11*, 4760 (1972)
207. Yeh, L.-S. L., Ingraham, L. L. in: Flavins and Flavoproteins. Singer, T. P. (ed.), p. 765. Amsterdam: Elsevier 1976
208. Dudley, K. H., Hemmerich, P.: Helv. Chim. Acta *50*, 355 (1967)
209. Brüstlein, M., Knappe, W.-R., Hemmerich, P.: Angew. Chem. *83*, 854 (1971)
210. Müller, F. et al.: Eur. J. Biochem. *19*, 392 (1969)
211. Massey, V., Curti, B., Müller, F., Mayhew, S. G.: J. Biol. Chem. *243*, 1329 (1968)
212. Müller, F., Szczesna, V.: unpublished results
213. Mager, H. I. X., Berends, W.: Rec. Trav. Chim. *84*, 1329 (1965)
214. Mager, H. I. X., Berends, W.: Rec. Trav. Chim. *91*, 630 (1972)
215. Hemmerich, P. in: 19th Mosbacher Colloquium, p. 249. Heidelberg: Springer Verlag 1968

216. Hemmerich, P., Müller, F.: Ann. New York Acad. Sci. *212*, 13 (1973)
217. Massey, V., Palmer, G., Ballou, D. in: Oxydases and Related Redox Systems. King, T. E., Mason, H. S., Morrison, M. (eds.), p. 25. Baltimore: University Park Press 1973
218. Spector, T., Massey, V.: J. Biol. Chem. *247*, 5632 (1972)
219. Entsch, B., Ballou, D., Massey, V.: J. Biol. Chem. *251*, 2550 (1976)
220. Hastings, J. W. et al.: Proc. Natl. Acad. Sci. USA *70*, 3468 (1973)
221. Ghisla, S. et al.: Proc. Natl. Acad. Sci. USA *75*, 5860 (1978)
222. Van Schagen, C. G., Grande, H. J., Müller, F.: Rec. Trav. Chim. *97*, 179 (1978)
223. Müller, F., Grande, H. J., Jarbandhan, T. in: Flavins and Flavoproteins. Singer, T. P. (ed.), p. 38. Amsterdam: Elsevier 1976
224. Müller, F.: unpublished data
225. Mager, H. I. X. in: Flavins and Flavoproteins. Yagi, K., Yamano, T. (eds.), p. 455. Tokyo: Japan Scientific Societies Press 1980
226. Orf, H. W., Dolphin, D.: Proc. Natl. Acad. Sci. USA *71*, 2646 (1974)
227. Hamilton, G. A.: Prog. Bioorg. Chem. *1*, 83 (1971)
228. Muto, S., Bruice, T. C.: J. Amer. Chem. Soc. *102*, 4472 (1980)
229. Ball, S., Bruice, T. C.: J. Amer. Chem. Soc. *102*, 6498 (1980)
230. Muto, S., Bruice, T. C.: J. Amer. Chem. Soc. *102*, 7559 (1980)
231. Iwata, M. et al.: J. Amer. Chem. Soc. *102*, 5036 (1980)
232. Franken, H.-D., Rüterjans, H. Müller, F.: unpublished results
233. Goddard, W. A., Harding, L. B. in: Biochemical and Clinical Aspects of Oxygen. Caughey, W. S. (ed.), p. 513. New York: Academic Press 1979
234. Gupta, N., Vennesland, B.: J. Biol. Chem. *239*, 3787 (1964)
235. Ostrowski, W., Zak, Z., Krawczyk, A.: Acta Biochem. Biophys. *15*, 241 (1968)
236. Massey, V. et al.: Biochem. Biophys. Res. Commun. *36*, 891 (1969)
237. Singer, T. P., Edmondson, D. E. in: Molecular Oxygen in Biology. Hayaishi, O. (ed.), p. 315. Amsterdam: North Holland 1974
238. Hinkson, J. W., Bulen, W. A.: J. Biol. Chem. *242*, 3345 (1967)
239. Steenkamp, D. J., Mallinson, J.: Biochem. Biophys. Acta *429*, 705 (1976)
240. Singer, T. P. et al., in: Flavins and Flavoproteins. Yagi, K., Yamano, T. (eds.), p. 277. Tokyo: Japan Scientific Societies Press 1980
241. Hall, C. L., Lambeth, J. D. in: Flavin and Flavoproteins. Yagi, K., Yamano, T. (eds.), p. 657. Tokyo: Japan Scientific Societies Press 1980
242. Barber, M. J. et al.: Arch. Biochem. Biophys. *201*, 468 (1980)
243. Siegel, L. M., Faeder, E. J., Kamin, H.: Z. Naturforsch. *27b*, 1087 (1972)
244. Faeder, E. J., Davis, P. S., Siegel, L. M.: J. Biol. Chem. *249*, 1599 (1974)
245. Vermilion, J. L., Coon, M. J.: J. Biol. Chem. *253*, 8812 (1978)
246. Lowe, H. J., Clark, W. M.: J. Biol. Chem. *221*, 983 (1956)
247. Mayhew, S. G.: Biochem. Biophys. Acta *235*, 276 (1971)
248. Mayhew, S. G., Massey, V.: J. Biol. Chem. *244*, 803 (1969)
249. Edmondson, D. E., Tollin, G.: Biochemistry *10*, 124 (1971)
250. Dubourdieu, M., Legall, J., Favaudon, V.: Biochem. Biophys. Acta *376*, 519 (1975)
251. Vetter, H., Knappe, J.: Hoppe-Seyler's Z. Physiol. Chem. *352*, 433 (1971)
252. Van Lin, B., Bothe, H.: Arch. Mikrobiol. *82*, 155 (1972)
253. Andrews, L. et al.: Biochem. Biophys. Res. Commun. *55*, 1165 (1973)
254. Vestling, C.: Acta Chem. Scand. *9*, 1600 (1955)
255. Brunori, M. et al.: J. Biol. Chem. *246*, 3140 (1971)
256. Gomez-Moreno, C., Choy, M., Edmondson, D. E.: J. Biol. Chem. *254*, 7630 (1979)
257. Iyanagi, T., Makino, N., Mason, H.: Biochemistry *13*, 1701 (1974)
258. Siegel, L. M. et al., in: Flavins and Flavoproteins. Kamin, H. (ed.), p. 523. Baltimore: University Park Press 1971
259. Lambeth, J. D., Kamin, H.: J. Biol. Chem. *252*, 2908 (1977)
260. Ruzicka, F. J., Beinert, H.: J. Biol. Chem. *252*, 8440 (1977)
261. Iyanagi, T.: Biochemistry *16*, 2725 (1977)
262. Coles, C. J., Edmondson, D. E., Singer, T. P. in: Flavins and Flavoproteins. Yagi, K., Yamano, T. (eds.), p. 101. Tokyo: Japan Scientific Societies Press 1980
263. Van Dijk, C.: unpublished results

264. Wood, P. M.: FEBS Lett *44*, 22 (1974)
265. Georg, P. in: Oxidases and Related Redox Systems. King, T. E., Mason, H. S., Morrison, M. (eds.), p. 15. New York: Wiley 1965
266. Conrad, M.: J. Theor. Biol. *79*, 137 (1979)
267. Andersen, R. D. et al.: Proc. Natl. Acad. Sci. USA *69*, 3189 (1972)
268. Burnett, R. M. et al.: J. Biol. Chem. *249*, 4383 (1974)
269. Mayhew, S. G., Ludwig, M. L. in: The Enzymes. Boyer, P. D. (ed.), Vol. 12, 3rd edit., p. 57, New York: Academic Press 1975
270. Mayhew, S. G., Faust, G. P., Massey, V.: J. Biol. Chem. *244*, 803 (1969)
271. Van Schagen, C. G., Müller, F.: Eur. J. Biochem. 1981, in press
272. Hemmerich, P. in: Transport by Proteins. Blauer, G., Sund, H. (eds.), p. 123. Berlin: Walter de Gruyter & Co. 1978
273. Harzer, G., Ghisla, S. in: Chemistry and Biology of Pteridines. Kislink, E., Brown, H. (eds.), p. 37. Amsterdam: Elsevier North Holland 1979
274. Palmer, G., Mildvan, A. S. in: Structure and Function of Oxidation-Reduction Enzymes. Åkeson, Å., Ehrenberg, A. (eds.), p. 385. Oxford: Pergamon Press 1972
275. De Wit, J. L. et al.: Helv. Chim. Acta 1981, submitted
276. Van Schagen, C. G., Müller, F., Kaptein, R.: Biochemistry 1981, in press
277. Moonen, C. T. W., Müller, F.: Biochemistry 1981, in press
278. Adman, E. T., Biochem. Biophys. Acta 549, 107 (1979)
279. James, T. L., Edmondson, D. E., Husain, M.: Biochemistry *20*, 617 (1981)
280. Benecky, M. et al.: Biochemistry *18*, 3471 (1979)
281. Datta, P. K., Spiro, T. G.: Biochemistry *19*, 1590 (1980)
282. Irwin, R. M. et al.: Biochemistry *19*, 4639 (1980)
283. Datta, P. K. et al.: Biochem. Biophys. Acta *623*, 77 (1980)
283. Kitagawa, T. et al.: Biochemistry *18*, 1804 (1979)
285. Wahl, Ph. et al.: Eur. J. Biochem. *50*, 413 (1975)
286. Visser, A. J. W. G., Grande, H. J., Veeger, C.: Biophys. Chem. *12*, 35 (1980)
287. Song, P.-S.: Int. J. Quant. Chem. *3*, 303 (1969)
288. Teitell, M. F., Fox, J. L.: Int. J. Quant. Chem. *18*, 449 (1980)
289. Platenkamp, R. J., Palmer, M. H., Visser, A. J. W. G.: J. Mol. Structure *67*, 45 (1980)

Semiquinone Formation in Flavo- and Metalloflavoproteins

Dale E. Edmondson[1] **and Gordon Tollin**[2]

1 Department of Biochemistry, Emory University, School of Medicine, Atlanta, Georgia 30322 and
2 Department of Biochemistry, University of Arizona, Tucson, Arizona 85721, USA

Table of Contents

I Introduction

Since the initial observation of flavin radical species by Michaelis and coworkers [1,2], the involvement of flavins in one-electron oxidation-reduction processes in biological systems has occupied the attention of workers in the field of redox enzymology up to the present time. Flavin coenzymes occupy a unique role in biological oxidations in that they are capable of functioning in either one-electron or two-electron transfer reactions. Due to this amphibolic reactivity, they have been termed in a recent review [3] "to be at the crossroads" of biological redox processes.

The ongoing research into the structure and mechanism of flavoenzymes has been the subject of several recent excellent reviews [3-6]. The proceedings of six symposia held at intervals over the past 16 years [7-12] provide an overall perspective on the progress of flavoenzyme research over this time period. The intent of this article will be to focus directly on the chemical and physical properties of the semiquinone form of flavin coenzymes to the extent that current knowledge permits, from the point of view of both model system studies and from existing knowledge of their properties in flavoenzyme systems. For an in-depth treatment of flavin and flavoenzyme redox properties in which the oxidized and hydroquinone forme as well as the semiquinone form are discussed as related to their biological function, the reader is refered to the article by F. Müller in this volume.

II Properties of Neutral and Anionic Flavin Semiquinones in Model Systems

1 Absorption Spectral Properties

Two ionic forms of flavin semiquinones have been shown to exist in flavoenzymes: the neutral form and the anionic form [13]

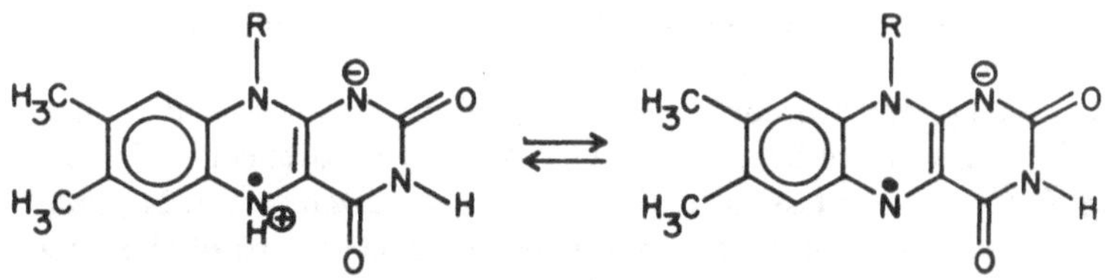

Neutral (blue) Anionic (red)

Cationic flavin semiquinones are formed below pH 3.0 [14]. Since this form is not found in biological systems, it will not be considered further.

The absorption spectral properties of the neutral and anionic forms are quite different as shown in Fig. 1. Due to the rapid dismutation of flavin radicals to form an equilibrium mixture with the hydroquinone and oxidized forms of the flavin, special procedures must be employed to measure the spectral properties of free flavin radicals. Nearly quantitative amounts of anion radical can be formed in aprotic solvents under basic conditions [15]. Alkylation of the N(5) position of the flavin hydroquinone followed by oxidation results in nearly quantitative formation of the

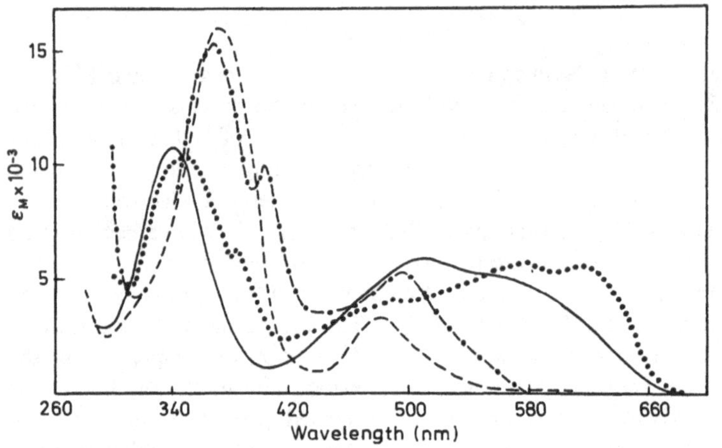

Fig. 1. Absorption spectra of neutral and anionic flavin semiquinones. (———) Anionic riboflavin semiquinone, (— — —) neutral riboflavin semiquinone [17]; (·········) neutral FMN semiquinone of *Azotobacter* flavodoxin, (—·—·—) anionic FAD semiquinone of glucose oxidase

neutral flavin semiquinone [16]. Absorption spectral properties of free anion and neutral flavin semiquinones in aqueous solutions have been measured using pulse radiolysis to generate the respective radicals and the absorption measured before decay by dismutation [17].

In contrast to their instability free in solution, flavin semiquinones bound to proteins are generally (although there are exceptions) quite stable and can be generated in nearly quantitative yields. This stability results from thermodynamic considerations in which the oxidized/semiquinone oxidation-reduction couple is usually more positive and well separated from the semiquinone/hydroquinone couple. In addition to thermodynamic stabilization, kinetic stabilization of protein-bound semiquinones has also been observed [18] and will be discussed in more detail in subsequent portions of this article.

Although the spectroscopic properties of the oxidized and hydroquinone forms of the flavin ring have been the subject of both theoretical [19-21] and experimental studies [22-25], much less attention has been given towards understanding the spectroscopy of the neutral and anion semiquinone forms of the flavin ring system. Perhaps the most definitive study to date on this subject is the comparison of polarized single crystal absorption spectral data on the oxidized and neutral semiquinone forms of *Clostridium* MP flavodoxin by Eaton and coworkers [26]. These studies show that the two lowest frequency $\pi - \pi^*$ transitions in the oxidized and neutral semiquinone forms have quite similar polarizations. Circular dichroism and absorption spectral data on solutions of the two oxidation-reduction forms of this flavodoxin also show that the two low frequency transitions exhibit similar anisotropy factors $g = \frac{\Delta\varepsilon}{\varepsilon}$ [21,27]. These data, as well as magnetic resonance data which indicate a qualitatively similar spin distribution for the triplet state of oxidized flavins and the doublet ground state of the semiquinone form [15,28-30] have led Eaton and coworkers to propose [26] that the electronic transitions of the neutral flavin semiquinone have the same orbital origins as the corresponding transitions of the oxidized form. This hypothesis would then imply that no gross electronic rearrangement in the flavin ring system occurs on one electron reduction. No corresponding single. crystal spectral data are yet available on the anionic flavin semiquinone. It is of interest, however, that the two low frequency transitions in the oxidized and anionic semiquinone forms of glucose oxidase (380 and 450 nm and 400 and 490 nm, respectively) have similar anisotropy factors ($g_{490}(Fl^{\overline{\cdot}}) = +1.6 \times 10^{-4}$ g_{450}Flox) $= 0.4 \times 10^{-4}$ and $g_{400}(Fl^{\overline{\cdot}}) = 5.4 \times 10^{-4}$, $g_{380}(Flox) = 4.9 \times 10^{-4}$) [27] which is suggestive that the same hypothesis applies regarding the relative electronic transitions of the oxidized and anionic semiquinone flavin species. Confirmatory experimental data must await single crystal polarization data on flavoproteins containing an anionic semiquinone form.

The model system studies of Müller et al. [31] have shown that alkylation of the O(2) and O(4) positions of the isolloxazine ring results in a flavin radical with similar ESR and absorption spectral properties as the anion flavin radicals although this flavin species has a neutral charge. Whereas binding of the flavin to its site on

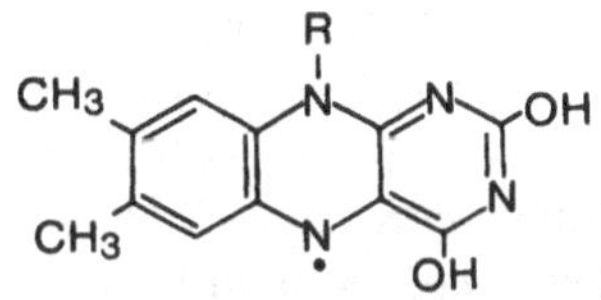

an apoprotein could, in principle, lead to this tautomeric neutral species which could result in anomalous chemical and spectral properties as compared to the "normal" neutral and anionic forms there has been, to our knowledge, no unequivocal demonstration of such a species in flavoenzymes. The possible existence of a neutral "red" flavin semiquinone in the flavoenzymes putrescine oxidase and Old Yellow Enzyme has been suggested [4] to explain the departure of these two enzymes from the correlation observed with a number of flavoenzymes between the formation of anionic semiquinones and a high affinity of the oxidized form to produce an N(5) sulfite adduct. These properties have been suggested to arise from the proximity of a protonated base of the protein to the N(1) position of the flavin ring resulting in an inductive effect which would facilitate sulfite addition to the N(5) position of the oxidized flavin and stabilize the anionic semiquinone. To demonstrate the existence of neutral red semiquinones in flavoenzymes, a great deal more must be learned about their chemical and physical properties so as to be able to distinguish them from the normal "anion" form.

2 Electron Spin Resonance (ESR) and Electron Nuclear Double Resonance (ENDOR) Spectral Properties

Since flavin semiquinones are tricyclic heteronuclear aromatic systems, the spin density due to the unpaired electron can be distributed at a number of sites on the isoalloxazine ring. Knowledge of the location and the extent of spin density at these locations on the ring system is important for the determination of sites that may participate in electron transfer reactions in enzymes. To accomplish this task, both theoretical and experimental approaches have been applied with the latter utilizing ESR and ENDOR spectroscopy.

It is essential to study the ESR spectrum of flavin radicals in solution since rapid tumbling results in the averaging out of anisotropic contributions of the hyperfine splitting and permits optimal resolution of the spectra. Flavin analogs with ribityl side chains (riboflavin, FMN, FAD) show complicated hyperfine splitting patterns since the side chain interacts strongly with solvent and does not undergo a rapid rotation. For this reason, most ESR studies have been done on lumiflavin analogs in which a methyl group replaces the ribityl side chain at the N(10) position.

The interpretation of the complex hyperfine splitting pattern of flavins was facilitated by comparing the ESR spectral properties of lumiflavin analogs of different isotopic and of different chemical group substitutions. The former method is preferable since isotopic substitution does not lead to any substantial alterations in electronic structure or geometry of the molecule. The results from studies of a large number of flavin analogs in their anionic [15,32)], neutral [16)], and cationic [33)] semiquinone forms provide a good understanding of the distribution of unpaired spin density about the flavin ring. Isotropic hyperfine coupling constants of nuclei at the various positions were evaluated from changes in the total linewidth of ESR spectra of flavin analogs differing in specific isotopic or chemical substitution. Computer simulations of spectra were beneficial in the analysis of neutral semiquinone [16)] and cationic semiquinone [33)] hyperfine coupling constants. The latter approach required the application of more sophisticated simulations utilizing Fourier transforms which proved to be successful, whereas earlier simulation approaches failed to adequately fit the experimental data for the estimation of the hyperfine coupling constants. ESR studies on fully deuterated FMN [34)] provided a means of directly measuring the coupling constants at the N(5) and N(10) positions, and at the exchangeable proton of N(5). These results were in good agreement with those of Müller et al. [16)].

Table 1. Isotropic hyperfine coupling constants of flavin semiquinones

Position	Nucleus	Anionic semiquinone[a] a(mT)[e]	Neutral semiquinone[b] a(mT)	Cationic semiquinone[c] a(mT)
1	N	—	—	—
3	N	—	—	—
5	N	0.73 ± 0.03	0.80 ± 0.02	0.732
	H(NH, NCH_3)	—	0.76 ± 0.02	0.779
6	H	0.35 ± 0.05	0.17 ± 0.02	0.134
7	H	—	—	0.050
8	H(CH_3)	0.40 ± 0.05	0.24 ± 0.05	0.322
9	H	0.09 ± 0.01	—	—
10	N	0.32 ± 0.03	0.36 ± 0.02	0.475
	H(NCH_3)	0.30 ± 0.02	0.39 ± 0.02	0.497
4a[d]	C	—(or+) 0.3	—(or+) 0.55	—(or+) 0.18

[a] Taken from reference 35. [b] Taken from reference 16. [c] Taken from reference 33. [d] Taken from reference 39. [e] 1mT = 10 gauss

Isotropic hyperfine coupling constants for the different nuclei of the flavin ring are given in Table 1 for the various ionic forms of the flavin semiquinone. An estimate of the unpaired spin density at each of the atoms exhibiting a detectable hyperfine coupling can be computed using the known values for the spin polarization parameters of the appropriate atom and the experimentally determined hyperfine coupling constant by means of the following expression:

$$a^x = Q_x P_x$$

where a_x is the hyperfine coupling constant of the atom x, Q_x is the spin polarization parameter of the atom, and P_x is the spin density in the P_z-orbitals of the respective carbon on nitrogen atoms. Detailed treatments of this relation have been published [36,38].

The results of these studies show that little or no spin density occurs in the pyrimidine ring, but rather the unpaired spin distribution is concentrated in the pyrazine and benzenoid portions of the isoalloxazine ring (Table 2). The site of highest spin density is the N(5) position with substantial densities also occuring at C(8) and at N(10). From ESR studies of 4a[1] ^{13}C enriched flavins, Walker et al. [39]

Table 2. Experimental and calculated spin densities of flavin semiquinones

Position	Nucleus	*Anionic semiquinone*		*Neutral semiquinone*	
		Experimental[a]	Calculated[b]	Experimental[c]	Calculated[d]
1	N		−0.002	~0	0.020
3	N		−0.036	~0	−0.021
5	N	0.256–0.394 (0.270)	0.361	0.282–0.432 (0.410)	0.382
6	C	0.129–0.167	0.201	0.067–0.085	0.146
7	C		−0.117	~0	−0.085
8	C	0.148–0.190	0.140	0.089–0.115	0.120
9	C	0.033–0.048	−0.060	~0	−0.077
10	N	0.112–0.173	0.164	0.126–0.194	0.328
4a[e]	C	0.126 ± 0.075 (+0.425)	0.163	0.023 ± 0.05 (+0.700)	—

[a] Taken from reference 35. [b] Taken from reference 37. [c] Taken from reference 16. [d] Taken from reference 40. [e] Taken from reference 39.

have shown that approximately 50% of the total spin density of cationic and anionic flavin radicals is located on the adjacent N(5) and C-4a nuclei.

A comparison of the various ionic forms suggests spin density at the C(4a) position to be lower for the neutral semiquinone that for the anionic or cationic forms [39]. These results are not unequivocal as it is not known what effect N(5) alkylation (required for stabilization of the neutral form) has on spin density at the C(4a)

[1] This position was designated C(12) in the original paper but is now referred to in the literature as the C(4a) position.

position. A diagrammatic representation of the sites of observed spin density on the flavin ring is shown in Fig. 2. It is therefore suggested that possible sites on the flavin ring important in one-electron transfers include the N(5), C(4a) and C(8) positions.

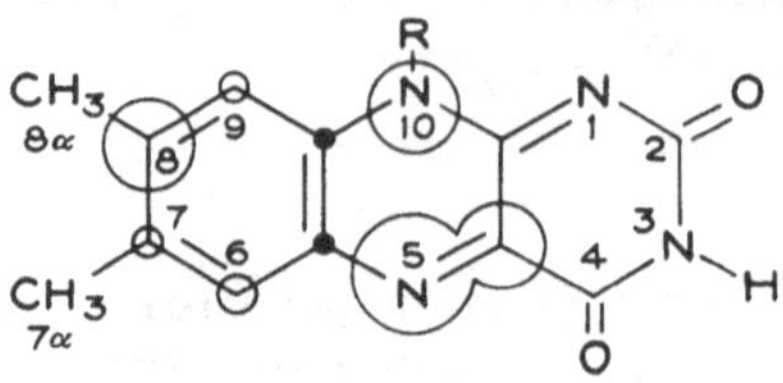

Fig. 2. Map of the flavin ring system and positions of spin densities. The size of the open circles approximates the relative amount of spin densities and the filled circles indicate sites of unknown spin density

While ESR approaches have proved invaluable in our understanding of the properties of flavin semiquinones in solution, information regarding protein-bound flavin radicals from ESR measurements are more limited. This is due to the slow tumbling on the ESR time scale which results in a "non-averaging" of anisotropic hyperfine coupling leading to the observation of broad, unresolved ESR spectra. Detailed properties of protein-bound semiquinones are of interest in determining whether or not binding to the apoprotein results in any substantial alterations in distribution of spin density on the flavin ring system. If any alterations are observed, such information would be useful in identifying a specific site(s) on the flavin ring important in electron transfer reactions involved in a specific flavoprotein mechanism.

ENDOR spectroscopy has proven to be a valuable technique to provide information on both free and protein bound flavin radicals. Since flavin radical ESR spectra can be partially saturated at moderate microwave power, ENDOR spectra may be observed as nuclear spin transitions by detection of changes in the partially saturated ESR signal as a function of nuclear radio frequency. The resonance condition for nuclei (when $I = {}^1/_2$) is described by the following equation:

$$\nu_i = |\nu_n \pm Ai/2|$$

where ν_N is the nuclear Zeeman frequency of the free nucleus and A_i is the hyperfine coupling in MHz. No proportionality between signal intensity and number of nuclei exists due to the dependence of the signal strength on the various longitudinal relaxation mechanisms.

Proton ENDOR studies on flavin and flavoprotein radicals have been published principally from Ehrenberg's laboratory [41,42]. Strong signals due to the 8-CH_3 protons are observed with couplings similar in value to the ones observed with the model systems (Fig. 3). Protein-bound anionic flavin semiquinones have 8-CH_3 isotropic couplings of 11 MHz where the neutral forms exhibit couplings of 7–8 MHz. These experiments show that the 8-CH_3 group in flavoproteins maintains isotropic character even at low temperatures and that binding of the flavin to a protein moiety has little influence on the hyperfine coupling to the 8-CH_3 protons.

The occurrence of shoulders on the ESR spectra of a number of flavoenzyme radicals is dependent on the temperature and microwave power used in the spectral experiments [43]. This anomalous saturation behavior is also observed for neutral

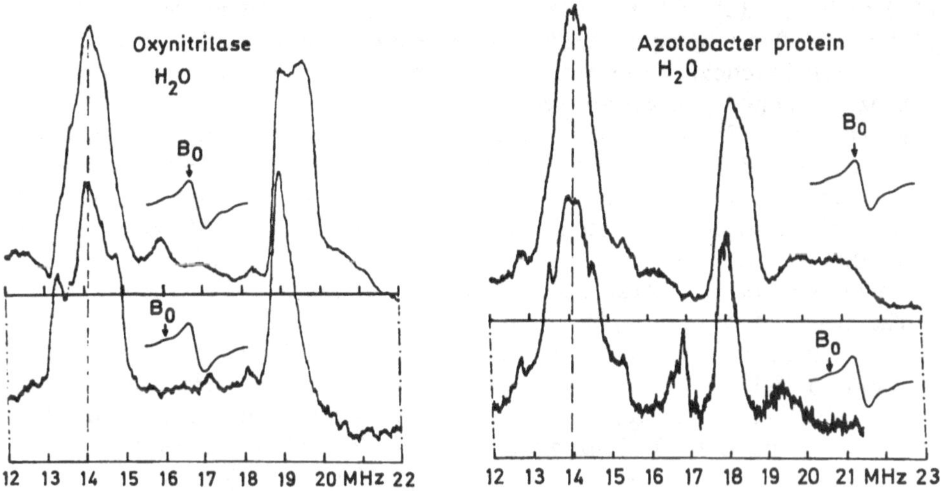

Fig. 3. ^{1}H-ENDOR spectra of a protein-bound anionic flavin semiquinone (oxynitrilase) and a protein-bound neutral flavin semiquinone (*Azotobacter* flavodoxin). The ENDOR spectra were recorded at the magnetic field settings indicated. Taken from Ref. [42] with permission

lumiflavin semiquinone in a toluene glass [44] at —120 °C but not in a polycrystalline phase at the same temperature. The origin of these shoulders has been suggested to be due to different saturation behavior for the (+1, +1) or (—1, —1) N(5) and N(10) nitrogen spin configurations. Values of 52 and 58 gauss are calculated for the anionic and neutral radicals respectively, based on the isotropic hyperfine coupling of these nuclei and on the assumption that the anisotropic interaction of the axial hyperfine tensors is 2.5 times that of the isotropic coupling [44]. Simulations of ESR spectra of totally deuterated flavin semiquinones [42] (in which the coupling from the N(5) and N(10) nitrogens predominate) provide a satisfactory fit of the experimentally obtained spectra (Fig. 4) [34]. Proton ENDOR spectra determined using the "wings" of the ESR spectrum for observation monitors only the components of the "powder" spectrum with fairly discrete orientations to the magnetic field. The observation of a narrow ENDOR signal due to the 8-CH_3 group in both anionic and neutral protein bound flavin semiquinones shows the orbital of

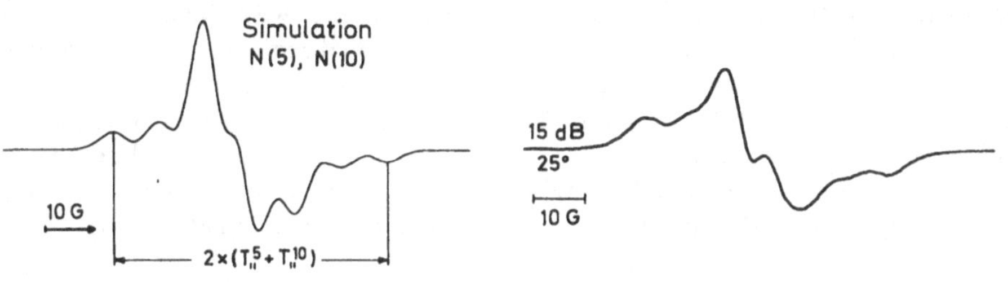

Fig. 4. Simulated ESR spectrum of a flavin semiquinone considering only the hyperfine interactions from the strongly coupled (N(5) and N(10) nitrogens (left) [42] and and experimentally observed ESR spectrum of a deuterated flavodoxin neutral semiquinone in ^2H_2O (right) [34]

C(8) to be parallel with the nitrogen orbitals, thus providing evidence for the planarity of flavin semiquinones [42]. More recent X-ray crystallographic data on the *Cl.* MP flavodoxin [45] provide confirmatory data on the planarity of the isoalloxazine ring in its one-electron reduced form.

Recent studies [46] have demonstrated the feasibility of solution ENDOR experiments on flavin cationic radicals and the observation of ^{14}N ENDOR signals from the strongly coupled N(5) and N(10) positions. The isotropic hyperfine coupling of both the nitrogen and protons (both + and —) are in good agreement with past ESR and ENDOR studies. Of potential interest for further work on the mechanism of flavin radical electron transfer reactions is the observation that replacement of N(5) by a sulphur atom results in a substantial redistribution of spin density in the benzenoid ring (i.e. a decrease at the 8 position and an increase at the 7 position). With the availability of more sensitive instrumentation and with increased radio frequency ranges and power capabilities, further applications of the ENDOR technique to flavoenzyme systems should lead to increased knowledge regarding the properties of flavin semiquinones in enzyme systems.

3 Interaction with Metal Ions

The rationale for studies on flavin semiquinone metal interactions stems from the presence of flavin coenzymes which participate in electron transfer in a number of metalloflavoproteins. Iron-containing redox centers such as the heme and non-heme iron sulfur prosthetic groups (Fe_2/S_2, Fe_4/S_4, or the "rubredoxin-type" of iron center) constitute the more common type of metal donor-acceptor found in metalloflavoproteins, although molybdenum is encountered in the molybdenum hydroxylases (e.g. xanthine oxidase, aldehyde dehydrogenase).

The metal chelating ability of flavins has been extensively investigated predominately in Hemmerich's laboratory [47]. While the oxidized and reduced (hydroquinone) forms of flavin are poor chelating ligands to divalent metal ions in aqueous solution, the flavin semiquinone forms chelates of considerable stability. This is readily demonstrated by the increase in semiquinone formation on the addition of metal ions to a half-reduced flavin solution which occurs by shifting the equilibrium away from semiquinone dismutation:

$$\mathrm{Flox} + \mathrm{FlH_2} \overset{-H^+}{\rightleftharpoons} 2\,\mathrm{Fl}^{\dot{-}} \overset{2Me^+}{\rightleftharpoons} 2\,\mathrm{Me}^+ - \mathrm{Fl}^{\dot{-}}$$

Stopped flow kinetic studies [48] show that metal ions react with the dimer of Flox and FlH_2 resulting in the formation of neutral and chelated semiquinone. Subsequent chelation of the neutral semiquinone by metal ions occurs at a much slower rate.

The stability constant for the lumiflavin semiquinone Zn chelate in aqueous solution was found to be $2 \times 10^4\ M^{-1}$, with the Cd chelate having about equal stability [49]. Aprotic solvents such as DMF serve to increase the stability constants relative to H_2O about 10-fold. In all systems tested, the stoichiometry of the flavin semiquinone-metal complex is 1:1.

ESR and ENDOR spectral data on flavin-metal chelates in which the metal has a magnetic moment (e.g. ^{67}Zn, ^{111}Cd, ^{113}Cd) have shown that the hyperfine couplings on the flavin are comparable to those of the anionic flavin semiquinone (as is the absorption spectra) and that the electronic spin is also delocalized to the metal ion [50]. These data suggest that the metal ion is coordinated to the N(5) and possibly the O(4) positions of the flavin ring. No ESR spectra are observed from flavin chelates with paramagnetic metal ions. Paramagnetic susceptibility measurements of a flavin semiquinone-Ni^{+2} complex showed the two spin systems ($S = {}^1/_2$ for the flavin; $S = 1$ for Ni^{+2} in a triplet state) to be coupled in a state with $S = {}^3/_2$ [50].

Recent X-ray and oxidation-reduction studies of a flavoquinone-Ru(II) complex [(10-methylisoalloxazine) · $(NH_3)_4Ru$] $(PF_6)_2 \cdot 2\,H_2O$ show the Ru(II) to be chelated to the flavin at the N(5) and O(4) positions. This metal complex is of interest in that Ru(II) is isoelectronic with low spin Fe [22] and thus serves as a model for Fe-flavin interactions. Of interest is the observation that, on complex formation, the flavin ring system deviates from co-planarity which, taken with absorption spectral and electrochemical data suggests a large amount of metal to ligand backdonation of electron density so that it may be viewed as a Ru(III)-Fl·complex. Whether deviations from non-planarity is general for flavin semiquinone-chelates will have to await further X-ray crystallographic investigations.

4 Oxidation-Reduction Properties

Thc redox and acid-base equilibria involving flavin semiquinone species are as shown in Scheme 1. pK^s values have been determined by both equilibrium (potentiometry) [52] and kinetic [17,53,54] methods and range from approximately 8.3 for flavins with uncharged R groups (lumiflavin, riboflavin) to 8.5 (FMN) and 8.8 (FAD) for phosphorylated flavins. The higher values for the coenzyme forms result from electrostatic repulsion between the negative charges in the isoalloxazine ring and the side chain phosphate.

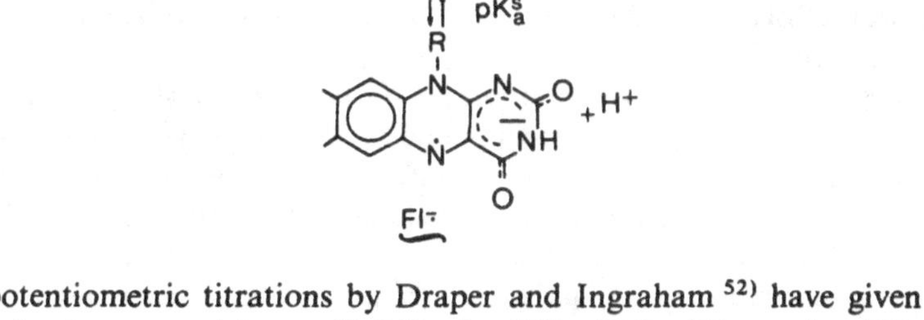

Careful potentiometric titrations by Draper and Ingraham [52] have given values (standard reduction potentials at pH 7.0) of −238 mV for E'_{02} and −172 mV for E'_{01} for FMN and corresponding values of −231 mV and −167 mV for riboflavin. Values of potential as a function of pH are tabulated in this paper and shown in a graphical presentation in Fig. 5.

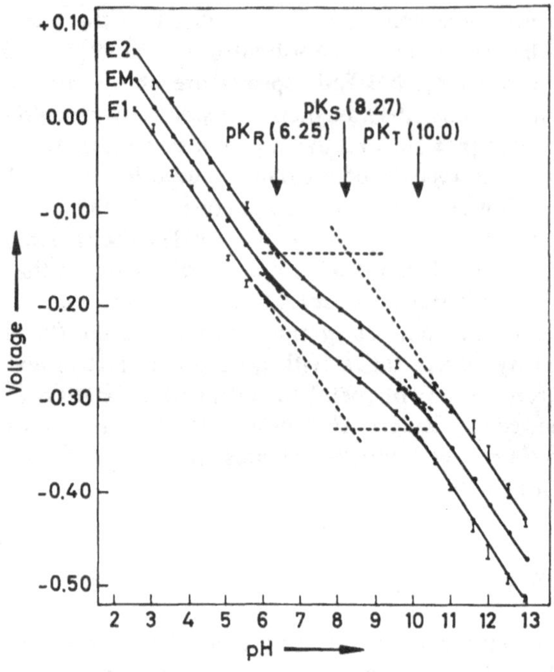

Fig. 5. The pH dependence of the three normal potentials of riboflavin. E_1 is the oxidized/semiquinone couple; E_2 is the semiquinone/hydroquinone couple; and E_M is the oxidized/hydroquinone (2-electron) couple. Taken from Ref. [54] with permission

5 Reaction with Oxidizing Agents

Flash photolysis [55] has proven to be a useful technique for investigating the reactivity of flavin radicals. This approach takes advantage of the fact that illumination of flavins in the presence of suitable reducing agents (e.g. EDTA, phenols, indoles) leads to a one-electron reduction of the flavin triplet state [53,55,56]. Through the use of a short duration flash, radical formation can be made to occur in several microseconds or less, and one can subsequently follow the fate of the semiquinone via optical absorption. By carrying out the photolytic conversion in the presence of oxidants, it is possible to directly measure electron transfer rate constants. Studies as a function of pH allow the reactions of both the neutral and anionic semiquinones to be monitored.

In Table 3 is a compilation of some of the rate constants which have been determined for various flavin semiquinone reactions. Note that whereas many of these reactions are quite rapid (at or near diffusion control), others are relatively slow. Of particular interest (see below) is the reaction of O_2 with FH·, which is too slow to measure (due to the competing disproportionation). Note also that side chain and electrostatic repulsion effects can be seen in some of these reactions.

A recent study [61] has shown that solvent dielectric constant exerts a considerable influence on flavin semiquinone reaction rates. A biphasic dependence was observed, with the rate constant being virtually independent of dielectric at low values and sharply increasing at high values. This was interpreted in terms of a change in

Table 3. Rate constants for flavin semiquinone reactions

Flavin semiquinone species*	Oxidant	k_2 ($M^{-1} s^{-1}$)	Comments	Ref.
LfH·	LfH·	6.2×10^9	disproportionation; pH = 5.0	[57]
RfH·	RfH·	1.1×10^9	pulse radiolysis; pH = 5.0	[17]
FMNH·	FMNH·	2.6×10^9	pH = 5.0	[57]
FADH·	FADH·	1.9×10^9	pH = 5.0	[57]
$Lf^{\overline{\cdot}}$	$Lf^{\overline{\cdot}}$	1.1×10^9	pH = 5.0	[57]
$Lf^{\overline{\cdot}}$	Lf^{+}	6×10^9	acetonitrile	[58]
3-MeLfH·	3-MeLf^{+}	4×10^9	aqueous buffer: pH = 7.0	[59]
LfH·	$Fe(CN)_6^{-3}$	6.0×10^8	pH = 5.0	[57]
$Lf^{\overline{\cdot}}$	$Fe(CN)_6^{-3}$	4.6×10^8	pH = 9.0	[57]
LfH·	cyt c	6.2×10^7	laser photolysis; pH = 7.0	[61]
LfH·	φO·	4.7×10^9	pH = 5.0	[57]
LfH·	O_2	$< 10^4$	upper limit only	[57]
$Lf^{\overline{\cdot}}$	O_2	3×10^8	flash photolysis	[57]
$Lf^{\overline{\cdot}}$	O_2	2.5×10^8	pulse radiolysis	[60]

* Abbreviations: Lf = lumiflavin; Rf = riboflavin; Me = methyl; φO· = phenol radical.

mechanism with solvent polarity. Specifically, it was suggested that hydrogen atom transfer and a neutral transition state occur in low dielectric media, and a dipolar transition state and electron transfer in high dielectric media. These results suggest the possibility that flavoenzyme apoproteins can control the rates and mechanisms of flavin semiquinone electron transfer reactions by providing the appropriate dielectric environment.

8α-Substitution of the flavin has been shown [62] to decrease the pK^s (by about 1 pH unit), to slightly decrease the rate of disproportionation (approximately a factor of two) and to decrease the rate of the reaction of $F^{\overline{\cdot}}$ with O_2 (2–10 times). Such effects are probably a consequence of electron withdrawal by the 8α-substituent and may have significance with respect to the enzymic properties of covalently bound flavin.

6 Interaction with Organic Compounds

Flavin, in all three of its oxidation states, readily forms molecular complexes with organic donors and acceptors [63]. Ingraham and coworkers [64,65] using potentiometric measurements, have determined affinity constants for the semiquinone form of riboflavin with several donors. These are in the range of several hundred M^{-1} for indoles, phenols and disulfides, all of which are known to be in the active sites of flavoproteins [66,67]. Such complex formation is undoubtedly of significance with respect to the effect of protein environment on flavin redox properties, especially semiquinone stabilization. In this context, Gillard and Tollin [68] have shown that complexation of FH· with tryptophan reduces the rate constants for several electron transfer reactions (O_2, ferricyanide, disproportionation) by as much as a factor of 10. It is interesting that tyrosine has a much smaller stabilizing effect even when complexation is enhanced by intramolecular effects [69].

7 Semiquinones as Intermediates in Flavin Redox Reactions

Inasmuch as flavins can accommodate two electrons but possess a relatively stable one-electron intermediate, an obvious question which can be asked of any flavin-mediated two electron redox reaction is whether or not the mechanism includes the radical species on a direct line between reactants and products. The mere observation of semiquinones in a reaction mixture is not sufficient evidence for their intermediacy, due to the existence of side reactions such as comproportionation ($F + FH_2 \rightleftharpoons 2\,FH\cdot$) which can generate radicals rapidly. Bruice [70] has discussed this question from a physical-organic point of view and concluded that there must exist a competition between one-electron and two-electron processes and that the actual mechanism should be determined mainly by the free energy of formation of substrate radical and the nucleophilicity of the substrate. Bruice has analyzed a variety of systems which he feels should proceed via one-electron mechanisms; among these are quinone and carbonyl group reduction by FH_2 [71].

Hemmerich and coworkers [72,73] have argued that a two-electron mechanism is most likely in reactions involving the formation and cleavage of carbon-hydrogen bonds, i.e. (de) hydrogenation. Unfortunately, no suitable chemical models are available for such reactions, although these workers have suggested that photochemical reactions of this type can be considered as analogous systems. On the basis of steady-state investigations of the flavin-sensitized oxidative photodecarboxylation of α-substituted acetic acids they propose a two-electron mechanism [72]. However, when such reactions are investigated using laser photolysis and spin-trapping techniques, clear evidence is obtained for a one-electron mechanism [74,75]. It should be realized that with an obligatory two-electron donor such as cyanoborohydride, flash experiments do indicate that a two-electron mechanism is operative [59], although it is noteworthy that the efficiency of this reaction, as measured by the second order rate constant for triplet reduction, is considerably less than for one-electron donors such as phenols or indoles, or for the α-substituted acetic acids.

The oxidation of FH_2 to F by O_2 has stimulated a good deal of interest since the pioneering work of Gibson and Hastings [76] which established the formation of the flavin radical during the reaction [77]. Although the mechanism of this reaction is quite complex, there is good reason to believe that a flavin semiquinone-superoxide ion pair is an intermediate [78]. A recent report [79] on the reaction of riboflavin semiquinone with O_2 has shown the existence of an intermediate with spectral properties similar to those of a 4a-peroxy reduced flavin which then further reacts to form Flox and H_2O_2. This is, to our knowledge, the first documentation of a flavin semiquinone oxidation mediated through a "complex" rather than through an "outer-sphere" type of electron transfer reaction.

III Flavin Semiquinones in "Simple" Flavoproteins

1 Electron-transferring Flavoproteins

a Flavodoxins

The flavodoxins are a group of FMN containing flavoproteins isolated from microorganisms which mediate electron transfer at a low redox potential between the prosthetic groups of other proteins (for reviews see references [66] and [80]). They function biochemically as one-electron carriers, shuttling between the semiquinone and fully-reduced forms, which has by far the most negative of the two potentials (see below). The ionization state of the semiquinone in the flavodoxins is FH· and the pK^s values are considerably higher than for free FMN [81].

The term "flavodoxin" was originally applied [82] to these flavoproteins because of their functional interchangeability with the ferredoxins, expressed as an induction of the synthesis of the flavodoxin as a replacement for ferredoxin under low iron growth conditions in organisms such as *Clostridium pasteurianum*. However, in other organisms (e.g. *Azotobacter vinelandii*) similar proteins, for which the same nomenclature is used, are constitutive [83] and play a unique biochemical role. It now appears that this type of low potential electron transfer flavoprotein is widely distributed among various types of microorganisms, including photosynthetic bacteria, blue-green and green algae, and nitrogen-fixing anaerobes and aerobes [66].

A large number of ferredoxin-dependent enzymes have been identified in bacteria [84]. Flavodoxin will replace ferredoxin in most of these. Among the biochemical processes in which these proteins function are nitrogen fixation, hydrogen production and sulfate reduction.

All known flavodoxins have low molecular weights (14,500–23,000 g $mole^{-1}$) and contain a single polypeptide chain and a single bound FMN. Upon the addition of one reducing equivalent, the FMN adds one electron and one proton to form the neutral semiquinone. This species is blue in color (as contrasted to the yellow color of the oxidized form) and has broad absorption bands in the visible between 400 and 700 nm. Addition of a second equivalent leads to a one-electron reduction to the hydroquinone, which is pale yellow in color and has weak bands at around 450 and 365 nm.

Magnetic resonance spectroscopy (ESR [43], ENDOR [42,44,85], NMR [86,87]) has been applied to the study of flavodoxins. The ESR and ENDOR results have indicated that the spin density distribution in flavodoxin semiquinone is similar to that found in free flavins and that the flavin radical is probably in a planar conformation. The latter conclusion is verified by X-ray structure work (see below). The NMR studies have argued against any extensive conformational changes upon change in redox state, have demonstrated that aromatic protons are shifted upfield when flavin is bound to apoprotein and disappear upon semiquinone formation, and have shown that the linewidths of some of the flavin resonances are consistent with rigid attachment of the flavin to the protein. All of these conclusions are in agreement with the X-ray crystallography.

The structure of two of the flavodoxins have been determined by X-ray crystal analysis. These are from *Clostridium MP* and *D. vulgaris*. Both of these flavodoxins contain a high degree of secondary structure and interesting similarities as well as differences are found. In the *Clostridium* protein [66,88,89] the FMN is bound on the periphery of the molecule. It is flanked by the side chains of Met 56 and Trp 90, the latter of which is directed toward the solvent side. Trp 90 and the FMN are inclined at an angle of 27° to one another. Only the dimethylbenzene portion of the flavin ring system is significantly exposed to solvent. It is through this region of the coenzyme that electron transfer has been suggested [66,81] to occur (see below).

The flavin-protein interactions serve to modulate the redox properties of FMN.

For *C.MP* flavodoxin at pH 7, the PFlH·/$PFlH_2$ and PFl/PFlH· redox couples are —399 mV and —92 mV respectively [66]. Examination of the structures of the flavodoxin semiquinone suggests how the change in the redox potential may be accomplished. In Scheme 1, it is seen that upon one electron reduction of the oxidized flavin to the semiquinone, position N(5) becomes protonated. In the structure of oxidized flavodoxin, no hydrogen bonding to this position is found. In the semiquinone structure, however, the hydrogen at N-5 forms a hydrogen bond to the carbonyl group of glycine 57. It is believed that residues 56 through 59 undergo a conformation change upon reduction from a distorted type II 3_{10} bend to a distorted type I bend. The formation of this hydrogen bond can partially account for the increased kinetic stability (see below) and the increased pK^s of the semiquinone, as well as explain the observed change in the redox potential. With regard to the latter, the 80 mV change in potential is equivalent to 1.8 kcal/mole, which is not an unreasonable value for the energy of a hydrogen bond. In electronic terms, the shift to a more positive value may be ascribed to a stabilization of the positive charge at N(5) via delocalization through the hydrogen bond. It is likely that the flavin-tryptophan interaction is also involved in semiquinone stabilization (see below).

For the *D. vulgaris* flavodoxin [90,91], although the location of the flavin, and the overall polypeptide folding patterns are remarkably similar to those for *Clostridium* flavodoxin the residues in the immediate vicinity of the flavin ring are different. Methionine is replaced by Trp 60, and tryptophan by Tyr 98. Tyr 98 is nearly coplanar with the isoalloxazine ring, whereas Trp 60 is held at about a 46° angle with respect to the flavin. These differences in the binding site environment are interesting in view of the similarities in the redox properties. It is not clear as yet whether semiquinone stabilization involves the same mechanisms as with the *Clostridium* protein.

The reaction of flavodoxin semiquinone with O_2 is quite slow (second order rate constants [92] for *A. vinelandii*, *C. pasteurianum* and *D. vulgaris* flavodoxins are 8.3×10^{-2}, 3.2 and 5.8 M^{-1}, s^{-1}, respectively). Although a rate constant for the reaction of free neutral flavin semiquinone (FH·) with oxygen has not been measured because of the competing disproportionation, the value for anionic form is 3×10^8 M^{-1} s^{-1} and the rate constant for FH· must be more than 10^4 times less than this (see above). Thus, the oxygen stability of flavodoxin semiquinones can be ascribed to this shift in the pK of the neutral form of the radical to higher values which prevents oxidation from occurring via the much more reactive anionic form. The disproportionation reaction is also effectively inhibited in the flavodoxins, possibly by steric effects, and thus the semiquinone is completely stable in the absence of oxygen. It is interesting that ferricyanide shows little or no difference in reactivity between the neutral and anionic free flavin radicals (see above). The slower rates of reaction of flavodoxin semiquinone, as compared with free flavin semiquinone, with this reagent (see below) must therefore be due to other factors, such as the flavin-tryptophan interaction.

The reaction of ferricyanide with the semiquinone forms of flavodoxins is more rapid than is the oxygen reaction. Second-order rate constants with *A. vinelandii*, *C. pasteurianum* and *D. vulgaris* flavodoxins are 8.3×10^3 M^{-1} s^{-1}, $1.1 \times 10^{5-1}$ s^{-1}, and 8.3×10^3 M^{-1} s^{-1} respectively [92]. This is to be compared with a value of

$6 \times 10^8\ M^{-1}\ s^{-1}$ for free flavin semiquinone (Table 3). The 7,8-dichloro-FMN analogs of *C. pasteurianum* and *A. vinelandii* flavodoxins react more slowly, with second-order rate constants of $3.0 \times 10^3\ M^{-1}\ s^{-1}$ and $1.4 \times 10^2\ M^{-1}\ s^{-1}$ respectively. Similar behavior has been observed with equine cytochrome c as an oxidant. Thus, for the three above mentioned flavodoxins, the measured[92)] rate constants are $1.1 \times 10^5\ M^{-1}\ s^{-1}$, $2.3 \times 10^5\ M^{-1}\ s^{-1}$ and $1.7 \times 10^{-1}\ s^{-1}$ for the normal proteins and $1.2 \times 10^3\ M^{-1}$ and $1.9 \times 10^3\ M^{-1}\ s^{-1}$ for the dichloro-FMN analogs of *A. vinelandii* and *C. pasteurianum* flavodoxins. For free flavin the value is $6.2 \times 10^7\ M^{-1}\ s^{-1}$ (see above). More recent work[93)] has shown that these flavodoxin reactions are somewhat more complex than was originally believed. A study of the concentration dependence of the observed rates has shown that a protein-oxidant complex is formed along the reaction pathway. However, decreases in the rate of electron transfer by chlorine substitution of more than two orders of magnitude are still obtained. This is thought to provide direct confirmation of the hypothesis that electron transfer, at least from the semiquinone form of the flavodoxin to iron-containing oxidants, proceeds via the dimethylbenzene portion of the flavin ring structure. It is interesting that model studies[94)] of the reaction of free flavins with cytochrome *c* indicate a similar participation of the dimethylbenzene ring. Recent ^{31}P NMR studies[95)] have shown that the rate of electron exchange between the flavin semiquinone and flavin hydroquinone of *M. elsdenii* flavodoxin is 100 sec^{-1} which is much greater than the exchange rate between the oxidized and semiquinone forms. In line with this observation are kinetic studies of dithionite reduction of *M. elsdenii* flavodoxin[96)] and temperature-jump kinetic studies of the methyl viologen radical equilibrium with *Azotobacter* flavodoxin[97)] which shows that the rate of conversion of the semiquinone to the hydroquinone to be at least two orders of magnitude faster than the rate for oxidized to semiquinone conversion. These data have been interpreted to suggest a change in protein conformation on reduction from the oxidized form to the semiquinone. Pulse radiolysis kinetic studies on Clostridium *MP* flavodoxin[98)] show that four spectrally distinct processes are involved in reduction of the oxidized FMN to the semiquinone by the hydrated electron. The slowest, last two species have been proposed to involve a protein conformational change on semiquinone formation. Recent and ongoing ^{31}P NMR studies on *Azotobacter* flavodoxin[99)] show chemical shift alterations of the two non-coenzyme phosphorus residues on semiquinone formation and on hydroquinone formation which suggests localized alterations in their environments consistent with alterations in protein conformation.

The above considerations provide a rationale for the redox properties of flavodoxin which function between the flavin hydroquinone and neutral semiquinone redox forms. Further studies are required to determine whether similar properties exist in flavoproteins in which both redox couples (PF/PFl· and PFl/$PFlH_2$) are operative and in situations where the anionic semiquinone rather than the neutral form is functional.

b Electron Transferring Flavoproteins (ETF)

The oxidation of fatty acids is catalyzed by the FAD-containing acyl coenzyme A dehydrogenases which transfer reducing equivalents to the mitochondrial respiratory chain *via* a flavin-containing electron transfer flavoprotein (ETF) and subsequently via an ETF dehydrogenase (an Fe/S flavoprotein[100)]. In addition to the mammalian form, ETF preparations have been isolated from an anaerobe [101] and from a methylotrophic bacterium [102]. The function of the ETF from the anaerobic bacterium *M. elsdenii* is to couple the oxidation of NADH to the reduction of butyryl coenzyme A dehydrogenase, a flavoprotein which catalyzes the reduction of short chain γ, β unsaturated fatty acids. The ETF isolated from the methylotrophic bacterium W3A1 [102] acts as the natural electron acceptor of the iron-sulfur, 6-S-cysteinyl FMN containing protein, trimethylamine dehydrogenase.

Each of the forms of ETF isolated from the different sources contain FAD as coenzyme and form an anionic semiquinone on one-electron reduction. Stopped-flow kinetic studies on the pig liver ETF [103] showed the anionic flavin semiquinone to be formed at times faster than catalytic turnover and thus demonstrate the participation of the anionic FAD semiquinone as an intermediate in the acceptance of reducing equivalents from the dehydrogenase. These studies would also imply the intermediacy of the semiquinone form of the acyl CoA dehydrogenase which would have been expected to form a neutral flavin semiquinone at the time the studies of Hall and Lambeth [103] were performed, however, no spectral evidence for its formation were found. Recent studies [104] have shown that the binding of CoA analogs to the dehydrogenase results in the perturbation of the pKa of the FAD semiquinone such that an anionic (red) rather than the neutral (blue) semiquinone is formed. This perturbation was estimated to reduce the pKa by at least 2.5 units to a value of 7.3. These data would suggest that both the dehydrogenase and ETF form anionic flavin semiquinone on electron transfer which cannot be differentiated spectrally, and thus the kinetic intermediacy of the anionic flavin semiquinone in electron transfer between the dehydrogenase and ETF may be due to both flavoproteins.

The properties of the semiquinone from of the ETF isolated from the methylotrophic bacterium [102] resemble those of the bacterial flavodoxins with the exception that flavodoxins form neutral semiquinones whereas this ETF forms an anionic semiquinone. Nearly quantitative anionic semiquinone formation is observed either in the presence of excess dithionite or when excess trimethylamine and a catalytic amount of trimethylamine dehydrogenase are added. Of interest is the apparent stability of the anionic semiquinone towards oxidation by O_2 [102] but not to oxidizing agents such as ferricyanide. This appears to be the first example of an air-stable protein-bound anionic flavin semiquinone. Future studies on the factors involved in imparting this resistance to O_2 oxidation by the apoprotein are looked forward to with great interest.

2 Pyridine Nucleotide-Dependent Flavoproteins

In the transfer of reducing equivalents from the pyridine nucleotide pool, flavoproteins carry out a central role of mediating the conversion of the obligatory 2-electron reductant to 1-electron receptors such as hemes and iron-sulfur redox centers. In such a role, the semiquinone form of the flavin serves as a pivotal intermediate. The reduction of flavins and flavoproteins by reduced pyridine nucleotides has been extensively studied since the initial work of Singer and Kearney [105] which showed that flavin reduction can occur in a non-enzyme catalyzed manner. The reduction proceeds as a 2-electron process since the formation of a nicotinamide semiquinone (a necessary intermediate in a 1-electron process) has been shown to be a thermodynamically unfavorable process (E_1 values in the range of —650 to —850 mV) [106].

Pyridine nucleotide-dependent flavoenzyme catalyzed reactions are known for the external monooxygenase [107] and the disulfide oxidoreductases [108]. However, no evidence for the direct participation of the flavin semiquinone as an intermediate in catalysis has been found in these systems. In contrast, flavin semiquinones are necessary intermediates in those pyridine nucleotide-dependent enzymes in which electron transfer from the flavin involves an obligate 1-electron acceptor such as a heme or an iron-sulfur center. Examples of such enzymes include NADPH-cytochrome P_{450} reductase, NADH-cytochrome b_5 reductase, ferredoxin —$NADP^+$ reductase, adrenodoxin reductase as well as more complex enzymes such as the mitochondrial NADH dehydrogenase and xanthine dehydrogenase.

Pyridine nucleotide-dependent reductases which contain only a single flavin such as a liver microsomal NADH-cytochrome b_5 reductase [109] and spinach ferredoxin —NADPH reductase [110] and adrenodoxin reductase [111] do not appear to form thermodynamically of kinetically stable semiquinones on reduction either by substrate or chemically with dithionite, even though flavin semiquinones must occur during catalytic turnover. This behavior, in retrospect, is not too surprising if we take into consideration the advantage of two equipotential successive one electron oxidations in reduction of two one-electron acceptors by the $PFl\cdot/PFlH_2$ and $PF/PFl\cdot$ couples. A large disparity in the potentials of the two couples could lead to the formation of a stable semiquinone intermediate which would be unable to be reduced by the obligate two-electron donor, i.e. reduced pyridine nucleotide. For the enzymes, NADH-cytochrome b_5 reductase and adrenodoxin reductase, the binding of NAD^+ and $NADP^+$, respectively, have been shown to increase the relative two electron redox couples by approximately 100 mV [109,111]. This has been interpreted [111] to provide the thermodynamic driving force for formation of fully reduced protein since the relative two-electron potentials of each enzyme (—291 mV [111]) and —258 mV [109] are close enough to that of the pyridine nucleotide couple (—316 mV) to expect that only partial reduction would occur. The oxidized forms of the species with bound pyridine nucleotide have little or no affinity for the respective oxidized forms of the two flavoenzymes. Such behavior is expected to provide a further driving force for complete reoxidation of the flavoenzymes by their electron acceptors. The addition of NAD^+ to a partially reduced solution of NADH-cytochrome b_5 reductase has been observed to facilitate comproportionation to form the anionic flavin semiquinone in a yield of 31% [109]. The catalytic significance of this observation is not known since no deviation from a two-electron Nernst plot is observed with NADH as reductant and no kinetic studies have been done to compare the rate of the NAD^+-facilitated comproportionation reaction with the rate of catalytic turnover. No comparable studies on the effect of $NADP^+$ on the oxidation-reduction potential of ferredoxin-$NADP^+$ reductase have been, to our knowledge, published. Inasmuch as the physiological role for this enzyme is *reduction* of the pyridine nucleotide rather than its oxidation, the potential of the enzyme should be significantly lower than that of the pyridine nucleotide couple. Indeed, a value of —445 mV has been determined for this flavoenzyme [112] with the driving force for its reduction being due to a decrease of 90 mV in the one-electron potential of the ferredoxin reductant. This increase in reducing power arises from the energy involves from the binding of the reduced ferredoxin to the oxidized flavoprotein reductase [112].

In contrast to the above reductases which contain only a single flavin moiety, NADPH-cytochrome P_{450} reductase contains both FMN and FAD [113]. One flavin may be presumed to accept two reducing equivalents from NADPH while the other serves as a one-electron reductant for the heme iron of cytochrome P_{450}. Flavin resolution and reconstitution studies [114] have shown that the FAD moiety is the low-potential flavin that accepts reducing equivalents from NADPH while the FMN moiety is the high potential flavin that serves as the one-electron reductant for cytochrome P_{450}.

A comprehensive series of oxidation-reduction potential measurements [113] have shown the FAD moiety to have the following one-electron couples: PFl/PFlH· = = —290 mV and PFlH·/$PFlH_2$ = —365 mV while the FMN moiety exhibits the following: PFl/PFl· = —110 mV and PFlH·/$PFlH_2$ = —270 mV. The FMN and FAD smiquinones were found to both be the neutral form as judged from absorption and ESR spectral data. The overlap of oxidized/semiquinone potential of the FAD moiety with the semiquinone/hydroquinone couple of the FMN moiety demonstrates the thermodynamic facilitation of flavin-flavin electron transfer *via* a one-electron mechanism. Stopped-flow kinetic data are also consistent with this view [115,116]. In contrast to NADH-cytochrome b_5 reductase and adrenodoxin reductase, the presence of oxidized pyridine nucleotide does not appear to influence the redox potential of either flavin of the reductase [113]. The effect of complexation of the reductase with cytochrome P_{450} [117] on the redox potentials of the individual flavin moieties or on the cytochrome have not, to date, been investigated. The binding of the reductase and electron transfer to the cytochrome is dependent both on phospholipid, and on a hydrophobic "tail" which is present in the detergent-solubilized reductase but not in the proteolytically solubilized form [117]. From a thermodynamic point of view, it would appear that P_{450} reduction is mediated between the "one-" and "three-electron" reduced forms of the reductase. This suggestion results from the known one-electron potential of the cytochrome (—237 mV in the presence of benzphetamine as substrate) [118] and the PFlH·/$PFlH_2$ couple of —270 mV for the flavin moiety of the reductase. The oxidized/semiquinone couple of the FMN (—110 mV) is clearly too positive to be a viable reductant although this suggestion is equivocal since the potential for the substrate-bound, ferrous oxy form of the cytochrome is not known and may, in fact, be more positive than the substrate-bound ferric form [119].

3 Enzyme-Bound Flavin Semiquinone which are not Catalytic Intermediates

A considerable amount of information regarding flavin semiquinone reactivity as well as the environment of the bound flavin coenzyme has accumulated over the years from studies of flavoenzyme systems which produce semiquinones either on photochemical reduction [13,119] or upon reduction by one electron equivalent of dithionite, but which do not form a detectable semiquinone intermediate during catalytic turnover. For example, the correlation of anionic semiquinone formation and the ability to bind sulfite at the N(5) position in a number of flavoenzyme oxidases [120] as well as "benzoquinoid" stabilization when either 8-mercapto or 6-hydroxy flavin analogs are bound have led Massey and coworkers (4) to suggest the proximity of a protonated basic amino acid residue to the N(1) position of the flavin ring. This interaction would be expected to exert an inductive effect which would facilitate nucleophilic addition by sulfite to the N(5) position as well as stabilize the anionic form of the flavin semiquinone.

In the case of two flavoenzyme oxidase systems (glucose oxidase (18) and thiamine oxidase [121] where both oxidation-reduction potential and semiquinone quantitation values are available, semiquinone formation is viewed to be kinetically rather than thermodynamically stabilized. The respective one-electron redox couples (PFl/PFl· and PFl·/PFlH$_2$) are similar in value (from essential equality to a 50 mV differential) which would predict only very low levels of semiquinone (32% when both couples are identical) at equilibrium. However, near quantitative yields (90%) of semiquinone are observed either by photochemical reduction or by titration with dithionite which demonstrates a kinetic barrier for the reduction of the semiquinone to the hydroquinone form. The addition of a low potential one-electron oxidoreductant such as methyl viologen generally acts to circumvent this kinetic barrier and facilitate the rapid reduction of the semiquinone to the hydroquinone form.

Glucose oxidase is one of the few known flavoenzymes in which the semiquinone pKa is in a physiological range [13,18]. Both equilibrium and kinetic data show a pKa of approximately 7.3 which is 1.5 pH units *lower* than that of free FAD [52,53]. Protonation of the anionic semiquinone is very rapid, being complete in the dead time of the stopped flow apparatus [18]. One of the more interesting properties of both the neutral and anionic flavin semiquinones of glucose oxidase is their reactivity with O_2. Model studies with free flavin semiquinones have established that anionic semiquinones are at least 10^4 more reactive with O_2 than the neutral form (see above). This differential reactivity is not too surprising when one compares the relative oxidation-reduction potentials for $O_2/O_2^{\overline{\cdot}}$ (—330 mV [122]), Fl/Fl$^{\overline{\cdot}}$ (—325 mV at pH 10) and Fl/FlH· (—146 mV at pH 5 [52]). The pKa for protonation of $O_2^{\overline{\cdot}}$ is 4.45 [123] and thus, the potential dies not change in the pH range of 5–10. In the case of glucose oxidase, however, the reaction rate of the anionic flavin semiquinone with O_2 (Em = —200 mV [18]) is only three-fold greater than that of the neutral form (Em = —63 mV [18]) with O_2. The temperature dependence of the reaction rates for both forms differ and provide evidence for the direct reaction of the neutral form with O_2, rather than an apparent reaction of O_2 with low levels of the anionic form in equilibrium with the neutral form. What is not easily understood is the relatively facile rate of reduction of an oxidant (O_2) with a potential of —330 mV by a reductant with an apparent potential of only —64 mV. While more work needs to be done to resolve this dilemma, it is appropriate to point out that both the anionic and neutral semiquinone forms of glucose oxidase used in these experiments with O_2 were subject to kinetic stabilization and thus it is inappropriate to predict their respective reducing properties on the basis of thermodynamic (equilibrium) data. Since both forms of the glucose oxidase flavin semiquinone appear to be similarly kinetically stabilized, perhaps it is not unexpected that they exhibit similar reactivities with O_2.

IV Flavin Radicals in Metalloflavoproteins

The examples of flavin semiquinones in the flavoenzyme systems discussed thus far have involved situations in which flavin-metal one-electron transfer reactions occurred via *intermolecular* interactions. In this present section we will describe current knowledge regarding *intramolecular* one-electron transfer between flavins and metal-containing redox centers which occur in the same enzyme molecule. Direct studies of flavin semiquinone properties and reactivities become more formidable due to absorption spectral overlap with the other redox centers, which also exhibit visible absorbance which changes with redox level. In addition, for the majority of systems investigated, the rates of electron transfer among the redox centers present are much more rapid than the rates of catalytic turnover [124] so that one is left with the prospect of observing concentrations of flavin semiquinones in equilibrium with other redox centers with the relative extent of oxidation-reduction being dependent on their relative redox potentials.

1 Interaction with Molybdenum and Iron-Sulfur Centers

a Flavin-Containing Molybdenum Hydroxylases

The involvement of flavin semiquinones in a metalloflavoenzyme is probably best illustrated from a discussion of xanthine oxidase and/or dehydrogenase. This enzyme catalyzes the oxidation of xanthine to uric acid as well as a number of other reactions. There exists a vast literature on this enzyme and the interested reader is referred to several recent reviews [125-128] for extensive treatments. In addition to the FAD moiety, the enzyme contains one molybdenum and two distinct Fe_2/S_2 centers which are reduced and oxidized at rates consistent with catalytic turnover [129,130]. Available data unequivocally support the molybdenum center as the site for the entry of reducing equivalents from the substrate into the enzyme while the FAD moiety serves as the site for electron donation to O_2 in the case of the oxidase and NAD^+ in the case of the dehydrogenase. The FAD moiety is reduced *via* one-electron equivalents by either the molybdenum or by the reduced Fe_2/S_2 centers. The relative oxidation-reduction potentials for the various redox centers in both the dehydrogenase and the oxidase are given in Table 4 based on published data [131-134]. It should be pointed out that the values presented in Table 4 demonstrate the difficulties involved in poising the potential at one temperature

Table 4. Oxidation-reduction potentials for the redox centers of xanthine oxidase (mV)

Condition	Mo(VI)/Mo(V)	Mo(V)/(VI)	FAD/FADH·	FADH·/FADH$_2$	Fe/S I	Fe/s II
room temperature[a]	−300 (fast)	−300 (fast)	−242	−227	−300	−200
pH 7.2	−344 (slow)	−306 (slow)				
low temperature[b] pH 7.1	−355 (fast)	−356 (fast)	−280	−244	−300	−255

[a] Taken from Ref. [130]. [b] Taken from Ref. [131].

and then monitoring the relative ratios of oxidized and reduced species at the much lower temperatures required for ESR measurements. Recent ESR studies [134] show that the ionic form of the FAD semiquinone has a measurable pKa (8.8), as was found with glucose oxidase [13,18]. The available redox potential data (Table 4) illustrates the thermodynamic feasibility for reduction of the flavin by the other redox centers in the enzyme.

When O_2 is the electron acceptor, the reduction can occur in either two-electron steps with $FADH_2$ as reductant and H_2O_2 as the product or in a one-electron manner with $O_2^{\bar{\cdot}}$ as the product. In the latter case, the reduced form of the flavin could be either $FADH_2$ or FAD· Recent studies on the reaction of O_2 with reduced xanthine oxidase has shown that reoxidation of the six-electron reduced enzyme by O_2 proceeds initially with two sequential two-electron steps to form two moles of H_2O_2 and the two-electron reduced form of the enzyme. Oxidation of the two-electron reduced form by O_2 then proceeds via two-sequential one-electron steps to form two moles of $O_2^{\bar{\cdot}}$ and oxidized enzyme. The differential rate of $O_2^{\bar{\cdot}}$ release [135,136] is suggestive of one mole arising from the one-electron oxidation of $FADH_2$ to FAD· while the other $O_2^{\bar{\cdot}}$ arises from the slower O_2-oxidation of FAD· to FAD. If this suggestion is correct, the observed insensitivity of the slow step to pH would suggest that the neutral and anionic forms of the FAD semiquinone exhibit similar reactivities with O_2. In this respect, it would be of interest to measure the relative oxidation-reduction potentials of the two ionic forms of the FAD semiquinone.

b Mitochondrial Succinate Dehydrogenase

The covalent 8α-N(3)-histidyl FAD of mitochondrial succinate dehydrogenase functions as a two-electron acceptor in the oxidation of succinate to fumarate and as a one-electron donor in the reduction of the iron-sulfur centers of the enzyme. Recent ESR spectroscopic data [137,138] have shown the covalent flavin semiquinone to be anionic in the pH range of 6.1 to 9.1. Indeed, a survey of all known flavoproteins containing an 8α-substituted covalently bound flavin has resulted in the finding that their semiquinone forms are anionic in the physiological pH range [137]. Recent potentiometric studies as a function of pH [139] have suggested the flavin semiquinone to have a pKa of 8.0, with no observed alterations in ESR spectral properties. The potentiometric data, however, is dependent on the quantitation of flavin semiquinone concentrations in equilibrium with the iron sulfur centers at certain poised potentials in the presence of mediator dyes as well as on the assumption that such equilibria are not altered on freezing the solution for ESR spectral measurements. In view of the differences observed in flavin semiquinone content whether the enzyme is reduced by substrate (32–48% in the pH range of 6.1 to 9.1 [135]) or by poised potentials (7–16% in the same pH range [139]) it would appear that additional investigations are required for a definitive description of the influence of such conditions on the flavin semiquinone. Magnetic interactions between the flavin semiquinone and iron-sulfur cluster S-1 (an Fe_2S_2 center) have been observed [139] demonstrating the proximity of the two redox centers which are estimated to be separated by a distance of 12–18 Å [139]. From these data and from the relative redox potentials of the two substrate reducible iron sulfur centers (S-1 Em = 0 mV and S-5, Em = +60 mV [136]) it has been suggested that S-1 is the iron-sulfur center which functions as the electron acceptor of the reduced flavin.

c Trimethylamine Dehydrogenase

This bacterial enzyme, which is isolated from a methylotrophic bacterium, is of interest to flavin enzymologists in two respects. The first is the presence of an unusual covalently bound FMN moiety which has been identified as 6-S-cysteinyl FMN [140, 141] and is present in a single molar stoichiometry with an Fe_4/S_4 cluster [142] on the enzyme. In addition, it is the only known metalloflavoenzyme where the rate of molecular electron transfer between the two above redox centers is slow enough to be followed by kinetic techniques in the msec time range [143].

Both stopped-flow and rapid freeze quench kinetic techniques show that the substrate reduces the flavin to its hydroquinone form at a rate faster than catalytic turnover [143]. Reoxidation of the flavin hydroquinone by the oxidized Fe_4/S_4 center leads to formation of a unique spin-coupled species at a rate which appears to be rate limiting in catalysis. Formation of this requires the substrate since dithionite reduction leads to flavin hydroquinone formation and a rhombic ESR spectrum typical of a reduced iron-sulfur protein [143]. The appearance of such a spin-coupled flavin-iron sulfur species suggests the close proximity of the two redox centers and provides a valuable system for the study of flavin-iron sulfur interactions. The publication of further studies of this interesting system is looked forward to with great anticipation.

2 Heme-Flavin Semiquinone Interactions

The biological importance of one-electron transfer reactions between flavin and heme centers is demonstrated in considering the flavoenzymes NADH-cytochrome b_5 reductase and NADPH-cytochrome P_{450} reductase. In addition to these systems, a number of flavoproteins are known which contain both flavin and heme in the same oligomeric unit. Of these proteins, the best characterized are flavocytochrome b_2 or yeast L(+) lactate dehydrogenase [141] and *Chromatium vinosum* cytochrome c-552 [145]. Lactate dehydrogenase contains FMN and heme in a one-to-one stoichiometry [145] while the bacterial flavocytochrome contains one covalently bound 8α-S-cysteinyl FAD [146] and two hemes [147]. Until recently, our knowledge regarding the rates of electron transfer reactions between hemes and flavins was quite meager. With the application of laser flash photolysis techniques to these systems [94, 147] it is anticipated that substantial new information will be forthcoming.

Recent studies on the electron-transfer between a number of flavin analogs with equine cytochrome *c* shows that flavin semiquinone reduction of cytochrome *c* reflects a diffusion-controlled process with a second-order rate of $6 \times 10^7\ M^{-1}\ sec^{-1}$ [145]. No major differences in rate were observed with the ionic form (anion or neutral) of the flavin semiquinone. Comparison of the rates of reduction by a number of flavin analogs suggests that cytochrome interaction occurs through the N(5)-dimethylbenzene region of the isoalloxazine ring.

Flash-photolysis studies on the heme-flavin electron transfer rate of Chromatium cytochrome c_{552} [147] shows that reduced heme (there are two in the protein) transfers a single electron to the oxidized flavin to form the anionic semiquinone with a first-order rate constant of $>1.4 \times 10^6$ sec^{-1}. A similar study of a related protein from Chlorobium (cytochrome C_{553}) has determined a rate constant of 1.2×10^3 s^{-1} for one-electron transfer from flavin semiquinone to heme [148]. These rapid rates of electron transfer demonstrate that in hemoflavoenzymes, as in metalloflavoenzymes containing molybdenum and iron-sulfur centers, intramolecular electron transfer is much faster than catalytic turnover and that the distribution of reducing equivalents observed during stopped-flow or rapid quench kinetic measurements which are limited to the msec time of slower scale is determined by the relative redox potentials of the redox centers involved. Recent kinetic studies on the flavocytochrome L-(+)-lactate dehydrogenase have shown that the two-electron reduction of the flavin by the substrate is rate limiting [149,150] with subsequent rapid formation of the reduced heme and neutral flavin semiquinone. The formation of the three-electron reduced enzyme (reduced heme and flavin hydroquinone) is much slower than catalytic turnover since this form can only arise from intermolecular electron transfer.

V Conclusions and Present Status of the Field

It should be obvious from the literature discussed in this article that progress in our understanding of the properties of flavin semiquinones and their role in flavoenzyme catalysis has increased dramatically over the past twenty years. This has been due to the application of sophisticated chemical and physical approaches, as well as to an increase in the number and diversity of flavoenzymes which have been purified to homogeneity in quantities sufficient for extensive study.

In spite of this progress, the gaps in our knowledge of the molecular mechanisms of the participation of flavins in one-electron transfer reactions are enormous. Whether the reduction of flavins by "obligatory" two-electron donors occurs by a concerted two-electron process or by sequential one-electron transfers remains a matter of controversy and is a subject under current active investigation. It is hoped that this review will convince the reader of the usefulness and necessity of redox potential measurements in the understanding of electron transfer reactions in flavoenzymes. These type of measurements have become more numerous in recent years; however, more information of this type is needed. We have seen that the apoprotein environment can alter the one-electron potentials of their respective bound flavin coenzymes by several hundred millivolts, yet virtually nothing is known, on a molecular basis, of how this is achieved.

Finally, the precise sites on the isoalloxazine ring which participate in semiquinone oxidation reactions in flavoenzymes will depend on whether the protein-bound semiquinone exhibits the same distribution of spin density as in the free system or in fact differs. Current ENDOR data suggests that the spin density in the 8-position of protein-bound flavin semiquinones is similar to that of the free systems. Further work is required to monitor spin distribution in the N(5) and

N(10) positions of the bound semiquinone and to determine, systematically, if any correlations between redox potential, spin density distribution, and ionic form of the semiquinone exist.

VI Acknowledgements

The authors wish to acknowledge research support from the National Science Foundation (PCM-8100770 to D. E. Edmondson) and from the National Institutes of Health (AM-15057 to G. Tollin).

VII References

1. Michaelis, L., Schubert, M. P., Smythe, C. V.: J. Biol. Chem. *116*, 587 (1936)
2. Michaelis, L., Schwarzenbach, G. H.: ibid. *123*, 527 (1938)
3. Walsh, C.: Accts. Chem. Res. *13*, 148 (1980)
4. Massey, V., Hemmerich, P.: Biochem. Soc. Trans. *8*, 246 (1980)
5. Hemmerich, P.: Bio(in)organic views on flavin-dependent one-electron transfer. In: Advances in Chemistry series no. 162, bioinorg. chem. II. (Raymond, K. N. ed.), pp. 312–219. Amer. Chem. Soc. 1977
6. Merrill, H. H., Lambeth, J. D., Edmondson, D. E., McCormick, D. B.: Formation and mode of action of flavoproteins. In: Ann. Rev. Nutr. Vol. 1 (Darby, W. J., Broquist, H. P., Olson, R. E. eds.) pp. 281–317. Ann. Rev. Inc. 1981
7. Slater, E. C. (ed): Flavins and flavoproteins, BBA Libr. *8*, Amsterdam: Elsevier 1966
8. Yagi, K. (ed): Flavins and flavoproteins. Tokyo, Univ. of Tokyo Press 1968
9. Kamin, H. (ed): Flavins and flavoproteins. Baltimore, Univ. Park Press 1971
10. Singer, T. P. (ed): Flavins and flavoproteins. Amsterdam, Oxford, New York: Elsevier 1976
11. Yagi, K., Yamano, T. (eds.): Flavins and flavoproteins. Tokyo: Japan Scientific Societies Press. 1980
12. Massey, V., Williams, C. H. (eds.): Flavins and flavoproteins. New York: Elsevier/North Holland, Inc. 1982
13. Massey, V., Palmer, G.: Biochem. *5*, 3181 (1966)
14. Ehrenberg, A.: Acta Chem. Scand. *14*, 766 (1960)
15. Ehrenberg, A., Müller, F., Hemmerich, P.: Eur. J. Biochem. *2*, 286 (1967)
16. Müller, F., Hemmerich, P., Ehrenberg, A., Palmer, G., Massey, V.: ibid. *14*, 185 (1970)
17. Land, E. J., Swallow, A. J.: Biochem. *8*, 2117 (1969)
18. Stankovich, M. T., Schopfer, L. M., Massey, V.: J. Biol. Chem. *253*, 4971 (1978)
19. Fox, J. L., Laßerge, S. P., Nishimato, K., Forster, L. S.: Biochem. Biophys. Acta *136*, 554 (1967)
20. Song, P.-S.: J. Phys. Chem. *72*, 536 (1968)
21. Dixon, D. A., Lindner, D. L., Branchand, B., Lippscomb, W. N.: Biochem. *18*, 5770 (1979)
22. Johansson, L. B. A., Davidsson, A., Lindblom, G., Naqvi, K. R.: ibid. *18*, 4249 (1979)
23. Ghisla, S., Massey, V., Lhoste, J. M., Mayhew, S. G.: ibid. *13*, 589 (1974)
24. Edmondson, D. E., Tollin, G.: ibid. *10*, 113 (1971)
25. Tollin, G.: ibid. *7*, 1720 (1968)
26. Eaton, W. A., Hofrichter, J., Markinen, M. W., Anderson, R. D., Ludwig, M. L.: ibid. *14*, 2146 (1975)
27. Edmondson, D. E.: Ph. D. Dissertation, Univ. of Arizona 1970
28. Lhoste, J. M., Hang, A., Hemmerich, P.: Biochem. *5*, 3290 (1966)
29. Erickson, L. E. G., Ehrenberg, A.: Acta Chem. Scand. *18*, 1437 (1964)
30. Ehrenberg, A., Erickson, L. E. G., Hyde, J.: Electron-nuclear double resonance from radicals of flavins and flavoproteins. In: Flavins and flavoproteins. (Kamin, H. ed.) pp. 141–152. Baltimore, Univ. Park Press 1971

31. Müller, F., Brustlein, M., Hemmerich, P., Massey, V., Walker, W. H.: Eur. J. Biochem. *25*, 573 (1972)
32. Guzzo, A. V., Tollin, G.: Arch. Biochem. Biophys. *105*, 380 (1964)
33. Müller, F., Grande, H. J., Harding, L. J., Dunham, W. R., Visser, A. J. W. G., Reinders, J. H., Hemmerich, P., Ehrenberg, A.: Eur. J. Biochem. *116*, 17 (1981)
34. Crespi, H. L., Norris, J. R., Katz, J. J.: Biochim. Biophys. Acta *253*, 509 (1971)
35. Ehrenberg, A., Muller, F., Hemmerich, P.: Eur. J. Biochem. *2*, 386 (1967)
36. McConnell, H. M., Chesnut, D. B.: J. Phys. Chem. *28*, 107 (1958)
37. Beinert, H.: Flavins and flavoproteins, including nonsulfur proteins. In: Biol. Applications of Electron Spin Resonance (Swartz, H. M., Bolton, J. R., Borg, D. C. eds.) pp., 351–410, New York, Wiley Intersc. 1972
38. Bernal, J., Rieger, P. H., Fraenkel, G. K.: J. Chem. Phys. *37*, 1489 (1962)
39. Walker, W. H., Ehrenberg, A., Lhoste, J.: Biochim. Biophys. Acta *215*, 166 (1970)
40. Song, P. S.: Ann. N.Y. Acad. Sci. *158*, **4ob** (1969)
41. Eriksson, L. E. G., Hyde, J. S., Ehrenberg, A.: Biochim. Biophys. Acta *192*, 211 (1969)
42. Eriksson, L. E. G., Ehrenberg, A.: Biochim. Biophys. Acta *295*, 57 (1973)
43. Palmer, G., Müller, F., Massey, V.: Electron paramagnetic resonance studies on flavoprotein radicals. In: Flavins and flavoproteins (Kamin, H. ed.) pp. 123–137. Baltimore, Univ. Park Press 1971
44. Hyde, J. S., Eriksson, L. E. G., Ehrenberg, A.: Biochim. Biophys. Acta *222*, 688 (1970)
45. Smith, W. W., Ludwig, M. L., Pattridge, K. A., Tsernoglou, D., Petsko, G. A.: Crystallographic Studies of Flavodoxins: Some correlation between structure and redox potential. In: Frontiers of Biol. Energetics, Vol. II (Dutton, P. L., Leigh, J. S. Scarpa, A. eds.) pp. 957–964. New York, San Francisco, London, Academic Press 1978
46. Bock, M., Lubitz, W., Kurreck, H., Fenner, H., Grauert, R.: J. Am. Chem. Soc. *103*, 5567 (1981)
47. Hemmerich, P., Lauterwein, J.: The structure and reactivity of flavin-metal complexes. In: Inorg. biochem., Vol. 2 (Eichhorn, G. L. ed.) pp. 1168–1190. Amsterdam, Elsevier 1975
48. Favaudon, V., Lhoste, J. M.: Biochem. *14*, 4739 (1975)
49. Müller, F., Eriksson, L. E. G., Ehrenberg, A.: Eur. J. Biochem. *12*, 93 (1970)
50. Ehrenberg, A.: Flavin radical-metal chelates. In: Vitamins and hormones, Vol. 28 (Harris, R. S., Munson, P. L., Diezfalusy, E. eds.) pp. 489–503, New York, London, Academic Press 1970
51. Clarke, M. J., Dowling, M. G., Garafalo, A. R., Brennan, T. F.: J. Biol. Chem. *255* (1970)
52. Draper, R. D., Ingraham, L. L.: Arch. Biochem. Biophys. *139*, 265 (1970)
53. Vaish, S. P., Tollin, G.: J. Bioenerg. *1*, 181 (1970)
54. Meisel, D., Neta, P.: J. Phys. Chem. *79*, 2459 (1975)
55. Holmstrom, B.: Photochem. Photobiol. *3*, 97 (1964); Acta Phys. Polon. *26*, 419 (1964)
56. Tegmer, L., Holmstrom, B.: Photochem. Photobiol. *5*, 223 (1966)
57. Vaish, S. P., Tollin, G.: J. Bioenerg. *2*, 61 (1971)
58. Ballard, S. G., Mauzerall, D. C., Tollin, G.: J. Phys. Chem. *80*, 341 (1976)
59. Hemmerich, P., Knappe, W. R., Kramer, H. E. A., Traber, E.: Eur. J. Biochem. *104*, 511 (1980)
60. Faraggi, M., Hemmerich, P., Pecht, I.: FEBS Lett. *51*, 47 (1975)
61. Ahmad, I., Tollin, G.: Biochem. *20*, 5925 (1981)
62. Edmondson, D. E., Rizzuto, F., Tollin, G.: Photochem. Photobiol. *25*, 445 (1977)
63. Tollin, G.: Molecular complexes of flavin. In: Molecular associations in biology (Pullman, B. ed.) pp. 393–409. New York, Academic Press 1968
64. Yeh, L. S. L., Ingraham, L. L.: A study of lumiflavin-amino acid interactions and their relation to the stability of the semiquinone form of flavin. In: Flavins and flavoproteins (Singer, T. P. ed.) pp. 765–774, Amsterdam, Elsevier 1976
65. Draper, R. D., Ingraham, L. L.: Arch. Biochem. Biophys. *139*, 265 (1970)
66. Mayhew, S. G., Ludwig, M. L.: Flavodoxins and electron-transferring flavoproteins. In: The enzymes (Boyer, P. (ed.), Vol. 12, pp. 57–118, New York, San Francisco, London, Academic Press 1975
67. Massey, V., Veeger, C.: Biochim. Biophys. Acta *48*, 33 (1961)
68. Gillard, J. M., Tollin, G.: Biochem. Biophys. Res. Commun. *58*, 328 (1974)

69. McCormick, D. B., Falk, M. C., Rizzuto, F., Tollin, G.: Photochem. Photobiol. *22*, 175 (1975)
70. Bruice, T. C.: Some physical organic studies dealing with flavin catalysis. In: Flavin and flavoproteins (Singer, T. P. ed.) pp. 244–260. Amsterdam, Elsevier 1976
71. Bruice, T. C., Yano, Y.: J. Am. Chem. Soc. *97*, 5263 (1975)
72. Knappe, W. R., Hemmerich, P.: Liebigs Ann. Chem. 2037 (1976)
73. Traber, R., Werner, T., Schreiner, S., Kramer, H. E. A., Knappe, W. R., Hemmerich, P.: 1e⁻ and 2e⁻ Transfer in the photochem. of flavins. In: Flavins and flavoproteins (Yagi, K., Yamano, T. eds.) pp. 431–442. Tokyo, Japan Scientific Societies Press 1980
74. Novak, M., Miller, A., Bruice, T. C., Tollin, G.: J. Am. Chem. Soc. *102*, 1465 (1980)
75. Heelis, P. F., Phillips, G. O.: Photobiochem. Photobiophys. *1*, 63 (1979)
76. Gibson, Q. H., Hastings, J. W.: Biochem. J. *68*, 368 (1962)
77. Massey, V., Palmer, G., Ballou, D. P.: On the reaction of reduced flavins with molecular oxygen. In: Oxidases and related redox systems (King, T. E., Mason, H. S., Morrison, M. eds.) pp. 25–49. Baltimore, Univ. Park Press 1973
78. Kemel, C., Chan, T. W., Bruice, T. C.: J. Am. Chem. Soc. *99*, 7272 (1977)
79. Anderson, R. F.: Flavin-oxygen complex formed on the reaction of superoxide ions with flavo-semiquinone radicals. In: Flavins and flavoproteins (Massey, V., Williams, C. H. eds.) pp. 278–283, New York, Elsevier North Holland 1982
80. Simondsen, R., Tollin, G.: Mol. Cell Biochem. *33*, 13 (1980)
81. Edmondson, D. E., Tollin, G.: Biochem. *10*, 133 (1971)
82. Knight, E., Hardy, R. W. F.: J. Biol. Chem. *241*, 2752 (1966)
83. Hinkson, J. W., Bulen, W. A.: ibid. *242*, 3345 (1967)
84. Yoch, D. C., Valentine, R. C.: Ann. Rev. Microbiol. *26*, 139 (1972)
85. Fritz, J., Muller, F., Mayhew, S. G.: Helv. Chim. Acta *56*, 2250 (1973)
86. Crespi, H. L., Norris, J. R., Katz, J. J.: Nature *236*, 178 (1972)
87. James, T. L., Ludwig, M. L., Cohn, M.: Proc. Natl. Acad. Sci. USA *70*, 3292 (1973)
88. Burnett, R. M., Darling, G. D., Kendall, D. S., LeQuesne, M. E., Mayhew, S. G., Smith, W. W., Ludwig, M. L.: J. Biol. Chem. *249*, 4383 (1974)
89. Ludwig, M. L., Burnett, R. M., Darling, G. D., Jordon, S. R., Kendall, D. S., Smith, W. W.: The structure of *Clostridium MP* flavodoxin as a function of oxidation state: some comparisons of the FMN-binding sites in oxidized, semiquinone, and reduced forms. In: Flavins and flavoproteins (Singer, T. P. ed.) pp. 393–421. Amsterdam, Elsevier 1976
90. Watenpaugh, K. D., Sieker, L. C., Jensen, L. H., LeGall, J. Dubourdieu, M.: Proc. Nat. Acad. Sci. USA *69*, 3185 (1972)
91. Watenpaugh, K. D., Sieker, L. C., Jensen, L. H.: A crystallographic structural study of the oxidation states of *Desulfovibrio vulgaris* flavodoxin. In: Flavins and flavoproteins (Singer, T. P. ed.) pp. 405–410. Amsterdam, Elsevier 1976
92. Shiga, K., Tollin, G.: Studies on the mechanism of electron transfer in flavodoxins. In: Flavins and flavoproteins (Singer, T. P. ed.) pp. 422–433. Amsterdam, Elsevier 1976
93. Jung, J., Tollin, G.: Biochem. *20*, 5124, (1982)
94. Ahmad, I., Cusanovich, M. A., Tollin, G.: Proc. Natl. Acad. Sci. USA *78*, 6724 (1981)
95. Moonen, C. T. W., Muller, F.: Biochem. *21*, 408 (1982)
96. Mayhew, S. G., Massey, V.: Biochem. Biophys. Acta *315*, 181 (1973)
97. Barman, B. G., Tollin, G.: Biochem. *11*, 4755 (1972)
98. Faraggi, M., Klapper, M. H.: J. Biol. Chem. *254*, 8139 (1979)
99. Edmondson, D. E., James, T. L.: Physical and chemical studies on the FMN and non-flavin phosphate residues in *Azotobacter* flavodoxin. In: Flavins and flavoproteins (Massey, V., Williams, C. H. Eds.) New York, Elsevier North Holland 1982
100. Ruzika, F. J., Beinert, H.: J. Biol. Chem. *252*, 8840 (1977)
101. Whitfield, C., Mayhew, S. G.: ibid. *249*, 2801 (1974)
102. Steenkamp, D. J., Gallup, M.: ibid. *253*, 4086 (1978)
103. Hall, C. L., Lambeth, J. D. L.: ibid. *255*, 3591 (1980)
104. Mizzer, J. P., Thorpe, C.: Biochem. *20*, 4965 (1981)
105. Singer, T. P., Kearney, E. B.: J. Biol. Chem. *183*, 409 (1950)
106. Blankenhorn, G.: Eur. J. Biochem. *67*, 67 (1976)

107. Massey, V., Hemmerich, P.: Flavin and pteridine monooxygenases. In: The enzymes (Boyer, P. ed.) pp. 191–252, New York, San Francisco, London, Academic Press 1975
108. Williams, C. H.: Flavin-containing dehydrogenases, In: The enzymes (Boyer, P. D. ed.) pp. 89–173, New York, San Francisco, London, Academic Press 1976
109. Iyanagi, T.: Biochem. *16*, 2725 (1977)
110. Foust, G. P., Mayhew, S. G., Massey, V.: J. Biol. Chem. *244*, 964 (1976)
111. Lambeth, J. D., Kamin, H.: ibid. *251*, 4299 (1981)
112. Batie, C. J., Kamin, H.: ibid. *256*, 7756 (1981)
113. Iyanagi, T., Makino, N., Mason, H. S.: Biochem. *13*, 1701 (1974)
114. Vermillion, J. L., Coon, M. J.: J. Biol. Chem. *253*, 8812 (1978)
115. Iyanagi, T., Makino, R., Anan, F. K.: Biochem. *20*, 1722 (1981)
116. Yasukochi, Y., Peterson, J. A., Masters, B. S. S.: J. Biol. Chem. *254*, 7097 (1979)
117. Fench, J. S., Guengerich, F. P., Coon, M. J.: ibid. *255*, 4112 (1980)
118. Gunsalus, I. C., Meeks, J. R., Lipscomb, J. D.: Ann. N.Y. Acad. Sci. *212*, 107 (1973)
119. Massey, V., Hemmerich, P.: J. Biol. Chem. *252*, 5612 (1977)
120. Massey, V., Müller, F., Feldberg, R., Schuman, M., Sullivan, P. A., Howell, L. G., Mayhew, S. G., Matthews, R. G., Foust, G. P.: ibid. *244*, 3999 (1969)
121. Gomez-Moreno, C., Choy, M., Edmondson, D. E.: ibid. *254*, 7630 (1979)
122. Wood, P. M.: FEBS Lett. *44*, 22 (1974)
123. Czapski, G., Dorfman, L. M.: J. Chem. Phys. *68*, 1169 (1964)
124. Olson, J. S., Ballou, D. P., Palmer, G., Massey, V.: J. Biol. Chem. *249*, 4364 (1974)
125. Bray, R. C.: The reactions and the structures of molybdenum centers in enzymes. In: Advances in enzymology (Meister, A. ed.) Vol. 51, pp. 107–165, 1980
126. Bray, R. C.: Molybdenum iron-sulfur flavin hydroxylases and related enzymes. In: The enzymes (Boyer, P. ed.) Vol. 12, pp. 300–419, New York, San Francisco, London, Academic Press 1975
127. Massey, V.: Iron-sulfur flavoprotein hydroxylases. In: The iron-sulfur proteins (Lovenberg, W. ed.) Vol. 1, pp. 301–360, New York, Academic Press 1973
128. Olson, J., Palmer, G.: Concepts and approaches to the understanding of electron transfer processes in enzymes containing multiple redox centers. In: Molybdenum and molybdenum-containing enzymes (Coughlan, M. P. ed.) pp. 543–568, Oxford, Pergamon Press 1980
129. Bray, R. C., Palmer, G., Beinert, H.: J. Biol. Chem. *239*, 2667 (1964)
130. Edmondson, D. E., Ballou, D., Van Heuvelen, A., Palmer, G., Massey, V.: ibid. *248*, 6135 (1973)
131. Cammack, R., Barber, M. J., Bray, R. C.: Biochem. J. *157*, 469 (1976)
132. Barber, M. J., Bray, R. C., Cammack, R., Coughlan, M. P.: ibid. *163*, 279 (1977)
133. Porras, A. G., Palmer, G.: Room temperature potentiometric measurements on xanthine oxidase. In: Flavins and flavoproteins (Massey, V., Williams, C. H. eds.) pp. 810–820, New York, Amsterdam, Oxford, Elsevier/North Holland 1982
134. Barber, M. J., Siegel, L. M.: Proton and electron affinities and magnetic interactions associated with the molybdenum flavin, and iron-sulfur centers of milk xanthine oxidase. In: Flavins and flavoproteins (Massey, V., Williams, C. H. eds.) pp. 796–804, New York, Amsterdam, Oxford, Elsevier/North Holland 1982
135. Hille, R., Massey, V.: J. Biol. Chem. *256*, 9090 (1981)
136. Porras, A. G., Olson, J. S., Palmer, G. P.: ibid. *256*, 9096 (1981)
137. Edmondson, D. E., Ackrell, B. A. C., Kearney, E. B.: Arch. Biochem. Biophys. *208*, 69 (1981)
138. Ackrell, B. A. C., McIntire, W., Edmondson, D. E., Kearney, E. B.: The semiquinone form of succinate dehydrogenase and other 8α-substituted flavoenzymes. In: Flavins and flavoproteins (Massey, V., Williams, C. H. eds.) pp. 488–491, New York, Amsterdam, Oxford, Elsevier/North Holland 1982
139. Ohnishi, T., King, T. E., Salerno, J. C., Blum, H., Bowyer, J. R., Maida, T.: J. Biol. Chem. *256*, 5577 (1981)
140. Steenkamp, D. J., Kenney, W. C., Singer, T. P.: ibid. *253*, 2812 (1978)
141. Steenkamp, D. J., McIntire, W., Kenney, W. C.: ibid. *253*, 2818 (1978)
142. Hill, C. L., Steenkamp, D. J., Holm, R. H., Singer, T. P.: Proc. Natl. Acad. Sci. USA *74*, 547 (1977)

143. Steenkamp, D. J., Singer, T. P., Beinert, H.: Biochem. J. *169*, 361 (1978)
144. Labeyrie, F.: Interactions and electron transfer between the flavodehydrogenase and the cytochrome b_2 components of yeast flavocytochrome b_2, In: Flavins and flavoproteins (Massey, V., Williams, C. H. eds.) pp. 823–832, New York, Amsterdam, Oxford, Elsevier/North Holland 1982
145. Bartsch, R. G., Meyer, T. E., Robinson, A. B.: Complex C-type cytochromes with bound flavin. In: Structure and function of cytochromes (Okunuki, K., Kamen, M. D., Sekuzu, I. eds.) pp. 443–451, Baltimore, Univ. Park Press 1968
146. Kenney, W. C., Singer, T. P.: J. Biol. Chem. *252*, 4767 (1977)
147. Cusanovich, M. A., Tollin, G.: Biochem. *19*, 3343 (1980)
148. Tollin, G., Meyer, T. E., Cusanovich, M. A.: Biochem. in press
149. Pompon, D.: Eur. J. Biochem. *106*, 151 (1980)
150. Capeillere-Blandin, C., Barber, M. J., Bray, R. C.: Differences in electron transfer rates among prosthetic groups between two homologous flavocytochrome b_2 (L-lactate cytochrome c oxidoreductase) from different yeasts. In: Flavins and flavoproteins (Massey, V., Williams, C. H. eds.) pp. 838–843, New York, Amsterdam, Oxford, Elsevier/North Holland 1982

FSC
www.fsc.org
MIX
Papier | Fördert gute Waldnutzung
FSC® C083411

Zeitfracht Medien GmbH
Ferdinand-Jühlke-Straße 7
99095 Erfurt, Deutschland
produktsicherheit@kolibri360.de